*The Unimaginable Mathematics
of Borges' Library of Babel*

La Biblioteca de Babel   **Haber**

> By this art you may contemplate the varia-
> tion of the 23 letters...
> *The Anatomy of Melancholy*, part. 2, sect. II
> mem. IV.

El universo (que otros llaman la Biblioteca) se compone de un número indefinido, y tal vez infinito, de galerías hexagonales, con variadas / con diversas / con vastos pozos de ventilación en el medio, cercados por barandas bajísimas. Desde cualquier hexágono, se ven los pisos inferiores y superiores: interminablemente. La distribución de las galerías es invariable. Veinticinco anaqueles, a cinco largos anaqueles por lado, cubren todos los lados menos uno; su altura, que es la de los pisos, excede apenas la de un bibliotecario normal. La cara libre da a un angosto zaguán, que desemboca en otra galería, idéntica a la primera y a todas. A izquierda y a derecha del zaguán hay dos gabinetes minúsculos. Uno permite dormir de pie; otro, satisfacer las necesidades fecales. Por ahí pasa la escalera espiral, que se abisma y se eleva a lo remoto. En el zaguán hay un espejo, que fielmente duplica las apariencias. Los hombres suelen inferir de ese espejo que la Biblioteca no es infinita (si lo fuera realmente ¿a qué esa duplicación ilusoria?); yo prefiero soñar que las superficies bruñidas figuran y prometen al infinito... La luz procede de unas frutas esféricas que llevan el nombre de lámparas. Hay dos en cada hexágono: transversales. La luz que emiten es insuficiente, incesante.

Como todos los hombres de la Biblioteca, he peregrinado en busca de un libro, acaso del catálogo de catálogos; ahora que mis ojos casi no pueden descifrar lo que escribo, me preparo a morir a unas pocas leguas del hexágono en que nací. Muerto, no faltarán manos piadosas que me tiren por la baranda; mi sepultura será el aire insondable: mi cuerpo se hundirá largamente y se corromperá y disolverá en el viento engendrado por la caída, que es infinita. Yo afirmo que la

The Unimaginable Mathematics of Borges' Library of Babel

William Goldbloom Bloch

OXFORD
UNIVERSITY PRESS
2008

OXFORD
UNIVERSITY PRESS

Oxford University Press, Inc., publishes works that further
Oxford University's objective of excellence
in research, scholarship, and education.

Oxford New York
Auckland Cape Town Dar es Salaam Hong Kong Karachi
Kuala Lumpur Madrid Melbourne Mexico City Nairobi
New Delhi Shanghai Taipei Toronto

With offices in
Argentina Austria Brazil Chile Czech Republic France Greece
Guatemala Hungary Italy Japan Poland Portugal Singapore
South Korea Switzerland Thailand Turkey Ukraine Vietnam

Copyright © 2008 by Oxford University Press, Inc.

Published by Oxford University Press, Inc.
198 Madison Avenue, New York, New York 10016

www.oup.com

Oxford is a registered trademark of Oxford University Press

All rights reserved. No part of this publication may be reproduced,
stored in a retrieval system, or transmitted, in any form or by any means,
electronic, mechanical, photocopying, recording, or otherwise,
without the prior permission of Oxford University Press.

"The Library of Babel," from COLLECTED FICTIONS by Jorge Luis Borges, translated by Andrew
Hurley, copyright © 1998 by Maria Kodama; translation copyright © 1998 by Penguin Putnam Inc.
Used by permission of Viking Penguin, a division of Penguin Group (USA) Inc.

Library of Congress Cataloging-in-Publication Data
Bloch, William Goldbloom.
The unimaginable mathematics of Borges' Library
of Babel / William Goldbloom Bloch.
  p. cm.
Includes bibliographical references and index.
ISBN 978-0-19-533457-9
1. Borges, Jorge Luis, 1899–1986—Knowledge—Mathematics.
2. Mathematics and literature. 3. Mathematics—Philosophy. I. Title.
PQ7797.B635Z63438 2008
868—dc22   2008017271

Contents

We do not content ourselves with the life we have in ourselves and in our own being; we desire to live an imaginary life in the mind of others, and for this purpose we endeavor to shine. We labor unceasingly to adorn and preserve this imaginary existence and neglect the real.
—Blaise Pascal, *Pensées*, no. 147

First Page of the Autograph Manuscript of "La Biblioteca de Babel" ii

Acknowledgments vii

Preface xi

Introduction xvii

The Library of Babel 3

CHAPTER 1 Combinatorics: Contemplating Variations of the 23 Letters 11

CHAPTER 2 Information Theory: Cataloging the Collection 30

CHAPTER 3 Real Analysis: The Book of Sand 45

CHAPTER 4 Topology and Cosmology: The Universe (Which Others Call the Library) 57

CHAPTER 5 Geometry and Graph Theory: Ambiguity and Access 93

CHAPTER 6 More Combinatorics: Disorderings into Order   107

CHAPTER 7 A Homomorphism: Structure into Meaning   120

CHAPTER 8 Critical Points   126

CHAPTER 9 Openings   141

Appendix—Dissecting the 3-Sphere   148

Notations   157

Notes   159

Glossary   165

Annotated Suggested Readings   175

Bibliography   181

Index   187

Last Page of the Autograph Manuscript of "La Biblioteca de Babel"   193

Acknowledgments

Pigmæos gigātum humeris impositos plusquam ipsos gigantes videre.
 —Didacus Stella (Diego de Estella), *In sacrosanctum Jesu Christi Domini nostri Evangelium secundùm Lucam Enarrationum*

I say with Didacus Stella, a dwarf standing on the shoulders of a giant may see farther than a giant himself.
 —Robert Burton, *Anatomy of Melancholy*, "Democritus to the Reader"

IT IS A PLEASURE TO ACKNOWLEDGE THE MANY DEBTS OF gratitude I owe; indeed, so much so that it's difficult to affix a starting point. Rather arbitrarily, I'll begin with Joe Roberts, the professor who introduced me to the concept of elegance in mathematics via the study of combinatorics. Around the same time, I read Rudy Rucker's *Geometry, Relativity and the Fourth Dimension*, which contains a lovely exposition of Riemann's century-old idea that a universe could be both finite and limitless. Another well-deserved "thank you" to the unremembered friend who, many years ago, put a copy of *Labyrinths* into my hand.

Leaping to the present day, I thank Tricia Arnold for endowing the fellowship that enabled me to travel to Buenos Aires. In a similar vein, I thank Susanne Woods and Wheaton College for supporting this project with time, resources, and encouragement. Not surprisingly, two librarians, Martha Mitchell and TJ Sondermann, were extremely helpful in identifying and obtaining old books and journal articles linking mathematics and Borges. Another marvelous staff member at Wheaton, Kathy Rogers, consistently provided vital textual support.

ACKNOWLEDGMENTS

I am grateful to everyone in Buenos Aires who assisted me, most especially Fernando Palacio, cultural mediator and translator *por excelencia*. The Director of the National Library of Argentina, Silvio Maresca, and the Associate Director, Roberto Magliano, were kind enough to meet with me and do whatever was within their power to facilitate this project. Clara Bayá, webmaster and semiofficial translator for the National Library, provided invaluable aid in guiding me first around the building and then around various rules that turned out to be surprisingly pliant. An anonymous guard at the old National Library and an anonymous librarian at the Miguel Cané Municipal Library were both also willing to bend rules and show me parts of their respective buildings that are generally off-limits to the public. The librarian, who was delighted that someone from the United States cared enough about Borges to visit the Miguel Cané Municipal Library, informed me that Argentine civil servants can't bear to read "The Library of Babel." Apparently, they take the Kafkaesque qualities of the tale quite personally, viewing the story as an extended slap against their daily work-life and their organizational systems.

My colleagues from the Humanities, Michael Drout and Hector Medina, acted as a pushmi-pullyu—see Lofting, 77–85—in jump-starting the project, in devoting the time to read and comment on my manuscript, and to talk over many of its points with me. Drout also provided etymologies for me when necessary, encouraged me to create the word "slimber" out of "slim" and "limber," and reassured me whenever I feared that I was using too many infinitives.

Eric Denton coordinated the first group reading and offered salient suggestions and collegial encouragement duly leavened with cynicism. ("Bill, your book is neither fish nor fowl.")

Anni Baker, Bernard Bloch, Tom Brooks, Michael Chesla, Bev Clark, Betsey Dyer, Lisa Lebduska, Shelly Leibowitz, Shannon Miller, Laura Muller, Rolf Nelson, John Partridge, Joel Relihan, Dorothea Rockburne, Pamela Stafford, David Wulff, and Paul Zeitz read this book in manuscript form and provided worthy and meaningful feedback. Any errors or infelicities remaining are, of course, solely my own.

Domingo Ledezma helped me out by translating some thorny passages in the story and Doug Jungreis confirmed my intuitions about Hopf fibrations. Julio Ortega encouraged me and introduced me to Borges' widow, the remarkable Maria Kodama.

John Wronoski of Lame Duck Books in Cambridge, Massachusetts first let me hold Borges' autograph manuscript of "La biblioteca de Babel" in my shaking hands, and then kindly let me use images of it in this volume. (By the way, the manuscript is for sale for approximately $650,000. Prior to learning this, I never actually ached to be a multimillionaire, but now I hereby publicly promise that if this book sells over three million copies, I will cheerfully call Mr. Wronoski to negotiate a price.)

Throughout the process, my editor, Michael Penn, combined abiding wisdom, keen grammatical insight, calming patience, and sly humor. Working with him was a continuous pleasure. Stefano Imbert, the illustrator, did a marvelous job capturing the ambience of the Library. Other people associated with Oxford University Press who helped shape the final result are Ned Sears, Stephen Dodson, and Keith Faivre.

A number of readers caught errors and sent them to me, notably Thomas Dobrzeniecki, Thomas Drucker, and Brain Hayes. Mark Eichenlaub contributed a better algorithm for calculating the median of the set of prime numbers less than 10^{100}.

On a number of occasions, my mother-in-law and my parents gave generously of their time and energy by watching my young children, allowing me to devote myself to this work. Speaking of my children, Dylan always loved the "pokey things" in the illustrations and Levi was always willing to cheer me up with a cartwheel performance. Finally, my wife Ingrid tolerated my obsessions, disjunctions, and corporeal absences as I wrangled with various parts of this book. Her multiform support was, and continues to be, vital and cherished.

Thank you, one and all.

Preface

One feels right away that this is the kingdom of books. People working at the library commune with books, with the life reflected in them, and so become almost reflections of real-life human beings.
—Isaac Babel, "The Public Library"

"WHO IS THE INTENDED AUDIENCE FOR THIS WORK IN progress?" This question, asked almost apologetically by a friend, stumped me for only a fraction of a second. With the clarity and explosiveness usually reserved for a rare mathematical insight, the answer burst from me: *Umberto Eco!* Polymath, brilliant semiotician, editor of the journal *Variaciones Borges*, interpreter of "The Library of Babel," and a favorite author for many years—Eco struck me as the ideal reader of this writing. (And Umberto, I hope you do read and enjoy this, someday.)

Of the more than six billion people who are *not* Umberto Eco, I imagine that those who'd find this work appealing would share, to varying degrees, the following traits: a familiarity with and affinity for Borges' works, especially "The Library of Babel"; a nodding, perhaps cautious, acquaintance with the thought that mathematics might not be the root of all evil; and the habit of rereading sentences, paragraphs, and stories for sheer delight, as well for playing with the superpositions of layers of available meanings.

While it's possible to set up a straw man and use it to wonder which way of presenting information is "better,"

A Multi-Claused Sentence vs. A Picture of Overlapping Sets

I take the view that the approaches are complementary; they aren't two opponents locked into a zero-sum game for which one side must prevail. So, since part of my not-so-hidden agenda is to persuade those of a literary temperament that mathematics can be more than the "problem/solution" model of much rudimentary education, I present a Venn diagram that visually encapsulates the speculations of the previous paragraph (figure 1).

The intended audience is the intersection of the three different sets of character traits. Judging mainly from the steady sales of Borges' fiction, I have managed to convince myself that besides you (presumably), there are at least several hundred thousand people who fit this description.

If, however, an unimaginative education or a particularly unpleasant teacher left a lingering distaste for all things mathematical, I hope this book acts as a corrective. Mathematics can be creative, whimsical, and revelatory all at once. More to the point, as embodied in the different meanings of the word "analysis," it is simultaneously a process and an intellectual structure. Borges, a great imbiber of mathematics, seems to have understood this idea and instantiated it in many of his stories—most especially "The Library of Babel." His imagination works in, through, out, about, and all around logical strictures.

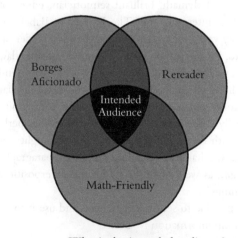

FIGURE 1. Who is the intended audience?

Conversely, for those of a mathematical bent who've not read Borges, I hope this volume inspires two things: a desire to explore more of Borges' work—there are many riches to be found—and, equally, a desire to learn more about the math tools I employ. We, as a society, are gifted these days; many books introducing math to the casual reader are readily available.

The chapters that are mathematical in nature will generally begin with the introduction of a mathematical idea. Some exposition, and perhaps a few examples, are given to help concretize the concept. Finally, the ideas will be applied to some aspects of "The Library of Babel" towards the desired end of producing an unimaginable (or unimagined) result.

Andrew Wiles, who proved Fermat's last theorem, memorably analogized the process of doing mathematics as follows:

> You enter the first room of the mansion and it's completely dark. You stumble around bumping into the furniture but gradually you learn where each piece of furniture is. Finally, after six months or so, you find the light switch, you turn it on, and suddenly it's all illuminated. You can see exactly where you were. Then you move into the next room and spend another six months in the dark.   (Singh, pp. 236–37)

Reading the math chapters of this work might be likened to stumbling around in a dark room, bumping into furniture, and finally, after finding the light switch, learning that you're not in a mansion after all, but rather facing away from the screen in a movie theater, and that the switch is really a fire alarm.

After the suite of introductory material comes the touchstone for this work: Andrew Hurley's superb translation of "The Library of Babel." After the story, and unlike most math books, the chapters are logically independent and can be dipped and skimmed as fancy dictates. (Of course, some intratextual references are unavoidable.) Although I've endeavored to structure the book so that it may be enjoyed from start to finish, based on predilections, nonlinear routes may be better suited for different kinds of readers.

In fact, it's safe to say that there are three main themes woven into this book. The first one digs into the Library, peels back layers uncovering nifty ideas, and then runs with them for a while. The second thread is found mostly in the "Math Aftermath" sections appended to the chapters: in them, I develop the mathematics behind the ideas to a greater degree and, in some cases, give step-by-step derivations for formulas used in the main body of the chapter. (Allow me to emphasize that the Math Aftermaths are—I hope—clear and engaging, but they certainly aren't required in order to understand and enjoy any other parts of the book.) The third focus is on literary aspects of the story and Borges; the chapters playing with these motifs come after those concerned with the math.

In the first chapter, "Combinatorics: Contemplating Variations of the 23 Letters," I use millennia-old ideas, alluded to in the story itself, to calculate the number of books in the Library. Once the basic concept of exponential notation is absorbed, the number is unexpectedly easy to find; it is understanding the magnitude of that number that occupies the bulk of the chapter. A number of previous critics also calculate this number, and several have provided similar means of understanding its size. By contrast, I fully explain the underlying mathematics and, moreover, add a new twist to the calculation. Expanding on some of the ideas raised, the Math Aftermath shows how to use a property of the logarithm function to recast the number of distinct books of the Library in terms more familiar, more amenable to our understanding. The chapter ends with the derivation of an ancient counting formula.

After that, in "Information Theory: Cataloging the Collection," I consider the meaning of a catalogue for the Library and the forms that it might take. The Math Aftermath takes some basic results in number theory and applies them to aspects of the Library and the unknowability of certain pieces of compressed information. Then, in "Real Analysis: The Book of Sand," I apply elegant ideas from the seventeenth century and counterintuitive ideas of the twentieth century to the "Book of Sand" described in the final footnote of the story. Three variations of the Book, springing from three different interpretations of the phrase "infinitely thin," are outlined.

Next, in "Topology and Cosmology: The Universe (Which Others Call the Library)," I employ late nineteenth- and early twentieth-century mathematics to explore possible shapes of the Library. Ultimately, I propose a rapprochement between the apparently conflicting views outlined by the narrator of the story. In the Math Aftermath section of the chapter, the discussion moves into somewhat more sophisticated domains by introducing two possible variations of the Library, each of which possesses noteworthy traits, one example being *nonorientability*.

Following this, in "Geometry and Graph Theory: Ambiguity and Access," I use Borges' descriptions of the Library to abstract the architecture of each floor of the Library and use it to unfold a surprising consequence. Interested readers can continue the tale of the chapter by following along in the Math Aftermath as I unpack an even stronger mathematical result stemming from the story.

The next chapter, "More Combinatorics: Disorderings into Order," is a kind of a fantasia on the possibilities inherent in ordering and disordering the distribution of books in the Library, and it concludes the mathematical section of the book.

After this, despite a desire to resist interpretation of the story, by drawing on metaphors from Alan Turing and information theory, I propose a new reading in "A Homomorphism: Structure into Meaning." Following that, in "Critical Points," prior work on "The Library of Babel" serves as a springboard to some compelling ruminations about life in the Library and other topics. Finally, in "Openings," a "What did he know and when did he know it? How did he know it?" attitude is adopted vis-à-vis Borges and mathematics. Was he a mathematician? A philosopher? A visionary writer blithely unaware of the depth of his insights?

The literary chapters are followed by a cortege of back matter, beginning with an appendix, "Dissecting the 3-Sphere," for those who want a refresher on how equations capture the characteristics and properties of multidimensional spheres. The appendix may sound scarier than it really is; I don't use much beyond the Pythagorean theorem, and I even provide a review of that.

In general, I avoid mathematical notation beyond that encountered in middle school or perhaps the early years of high school. However, in case it is unfamiliar, following the appendix is a short list of notations with definitions. Speaking of definitions, there's a lot to say on the matter. Mathematics is an intellectual discipline built on definitions; indeed, the *axioms* of mathematics are exactly definitions that have been accepted as plausible and true by the concerted critical faculty of millions of thinkers around the world aggregated over the past several millennia. Moreover, these days great theoretical breakthroughs occur when brilliant mathematicians see new interrelations and make definitions that enable a cascade of untold consequences to be discovered by other workers in the field. For us, definitions will be considerably more prosaic; I italicize words that strike me as being of a technical nature, outside the usual range of quotidian use, and provide definitions in a glossary following the notations and the endnotes.

As a reader, when I encounter an endnote, I'm compelled almost against my will to flip to the back of the book to learn what the endnote says.[1] As I writer, I find that despite my best efforts to incorporate them into the body of the book, my work includes diverting digressions, fine points of mathematics that might interest only specialists, and citations to other works. All of these are consigned to the endnotes.

After the glossary, an annotated list of suggested readings is provided for those with curiosity primed to learn more of the mathematics used in the book. A bibliography of references cited or consulted rounds out the end matter.

Introduction

We adore chaos because we love to produce order.

—M. C. Escher

It's an ironic joke that Borges would have appreciated: I am a mathematician who, lacking Spanish, perforce reads "The Library of Babel" in translation. Furthermore, although I bring several thousand years of theory to bear on the story, none of it is literary theory.

Having issued these caveats, it is my purpose to make explicit a number of mathematical ideas inherent in the story. My goal in this task is not to reduce the story in any capacity; rather it is to enrich and edify the reader by glossing the intellectual margins and substructures. Borges was a consummate synthesist; his lapidary prose sparkles and reveals unexpected depths when examined from any angle or perspective. I submit that because of his well-known affection for mathematics, exploring the story through the eyes of a mathematician is a dynamic, useful, and necessary addition to the body of Borgesian criticism.

In what follows, I assume no special mathematical knowledge. I only ask that the reader trust that I am a tour guide through a labyrinth, like that marble pathway on the floor of the cathedral at Chartres, not the gatekeeper of a Stygian maze without center or exit (figure 2). Beyond enhancing the story, the reader's reward will be an exposure to some intriguing and entrancing mathematical ideas.

FIGURE 2. The labyrinth on the floor of the Chartres Cathedral. Movement in a labyrinth is constrained to only forwards or backwards motion. (Jeff Saward/Labyrinthos)

Borges was a master of understating ideas, allowing them the possibility of gathering heft and power, of generating their own gravity. I'm under no delusion that he traced out all the consequences of the dormant mathematics I uncover. I allow myself the ambition, though, to paraphrase what Borges wrote in a forward and hope that this book would have taught him many things about himself (see Barrenchea, p. vii).

I request a last indulgence from the reader. The introductory material, thus far, has been written in the friendly and confiding first person singular voice. Starting in the next paragraph, I will inhabit the first person plural for the duration of the mathematical expositions. This should not be construed as a "royal we." It has been a construct of the community of mathematicians for centuries and it traditionally signifies two ideas: that "we" are all in consultation with each other through space and time, making use of each other's insights and ideas to advance the ongoing human project of mathematics, and that "we"—the author and reader—are together following the sequences of logical ideas that lead to inexorable, and sometimes poetic, conclusions.

A word, too, about the language in the book. We started our college years intending to be some sort of creative writer. Beyond the insight mathematics offered into the natural world and epiphenomena of life, and beyond the aesthetic joy at understanding how the iron rules of logic crystallize a good proof into a work of art, one of the reasons we turned to math was the lilt and rhythm of the "if-then" syntax coupled with the musicality of words often repeated, such as "thus," "hence," "suppose," and "let." We hope our readers might develop an ear for this music, too.

We close the introduction by offering several related disclaimers. Mathematics, like any discipline, is not a monolith; it's a sprawling agglutination of overlapping and intersecting fields and specialties: one's talents, tastes, and beliefs determine individual focus. We carefully checked and rechecked our ideas, mathematics, and figures. To the best of our knowledge, there are no mistakes. However, a different mathematician might well expose divergent mathematical themes from the story and utilize different sets of ideas to explain them.

Furthermore, there's a natural tendency for an individual reaching across traditional boundaries to be perceived as a universal embodiment of the foreign, the other. Although our inductions and deductions are correct, some mathematicians might issue philosophic challenges to underlying assumptions, especially in the chapters "Real Analysis" and "More Combinatorics." Consequently, no one, including the author, should be seen as a Representative or Ambassador, speaking in one voice for an ideologically unified Entity of Mathematicians: such an Entity of Mathematicians simply doesn't exist. (Lest this be subject to misinterpretation, allow us to note that *all* mathematicians would agree on the centrality of logically consistent deductions and derivations from agreed-upon axioms.)

It's important to bear in mind that the mathematical expositions contained herein are not rigorously developed, nor are they intended as comprehensive introductions to the various theories. Just as a stirring musical performance will not transform a concertgoer into a musician, composer, lyricist, musicologist, or music critic, so this book won't transform a reader into any kind of a mathematician. However, just as a concert may move, inspire, or transfigure a listener, so we hope that this book will stimulate, dazzle, and expand its readers.

Finally, about the title of the book: why the word "unimaginable"? By way of an answer, we note that in his sixth Meditation, Descartes makes clear the distinction between simply naming a thing and visualizing it in a clear, precise way that allows for mental manipulations.

> I note first the difference between imagination and pure intellection or conception. For example, when I imagine a triangle, I not only conceive it as a figure composed of three lines, but moreover consider these three lines as being present by the power and internal application of my mind, and that is properly what I call imagining. Now if I wish to think of a chiliagon, I indeed rightly conceive that it is a figure composed of a thousand sides, as easily as I conceive that a triangle is a figure composed of only three sides; but I cannot imagine the thousand sides of a chiliagon, as I do the three of a triangle, neither, so to speak, can I look upon them as present with the eyes of my mind.

Some of the ideas we'll talk about, such as titanic numbers and higher dimensions, are unimaginable in this sense. We can give names to the ideas, use metaphors to approach them, give simple examples to substitute in as models, and try to find a consistent set of rules and mathematical objects that encapsulate the essence of the ideas—but we will never be able to visualize them any more than we could Descartes' thousand-sided chiliagon. Indeed, our task as your guide is to trigger the processes by which you build intuition and insight into the Unimaginable.

*The Unimaginable Mathematics
of Borges' Library of Babel*

The Library of Babel
Jorge Luis Borges

> By this art you may contemplate the variation of the 23 letters....
> —*Anatomy of Melancholy*, Pt. 2, Sec. II, Mem. IV

THE UNIVERSE (WHICH OTHERS CALL THE Library) is composed of an indefinite, perhaps infinite number of hexagonal galleries. In the center of each gallery is a ventilation shaft, bounded by a low railing. From any hexagon one can see the floors above and below—one after another, endlessly. The arrangement of the galleries is always the same: Twenty bookshelves, five to each side, line four of the hexagon's six sides; the height of the bookshelves, floor to ceiling, is hardly greater than the height of a normal librarian. One of the hexagon's free sides opens onto a narrow sort of vestibule, which in turn opens onto another gallery, identical to the first—identical in fact to all. To the left and right of the vestibule are two tiny compartments. One is for sleeping, upright; the other, for satisfying one's physical necessities. Through this space, too, there passes a spiral staircase, which winds upward and downward into the remotest distance. In the vestibule there is a mirror, which faithfully duplicates appearances. Men often infer from this mirror that the Library is not infinite—if it were, what need would there be for that illusory replication? I prefer to dream that burnished surfaces are a figuration and promise of the infinite.... Light is provided by certain spherical fruits that bear the name "bulbs." There are two of these bulbs in each hexagon, set crosswise. The light they give is insufficient, and unceasing.

Like all the men of the Library, in my younger days I traveled; I have journeyed in quest of a book, perhaps the catalog of catalogs. Now that my eyes can hardly make out what I myself have written, I am preparing to die, a few leagues from the hexagon where I was born. When I am dead, compassionate hands will throw me over the railing; my tomb will be the unfathomable air, my body will sink for ages, and will decay and dissolve in the wind engendered by my fall, which shall be infinite. I declare that the Library is endless. Idealists argue that the hexagonal rooms are the necessary shape of absolute space, or at least of our *perception* of space. They argue that a triangular or pentagonal chamber is inconceivable. (Mystics claim that their ecstasies reveal to them a circular chamber containing an enormous circular book with a continuous spine that goes completely around the walls. But their testimony is suspect, their words obscure. That cyclical book is God.) Let it suffice for the moment that I repeat the classic dictum: *The Library is a sphere whose exact center is any hexagon and whose circumference is unattainable.*

Each wall of each hexagon is furnished with five bookshelves; each bookshelf holds thirty-two books identical in format; each book contains four hundred ten pages; each page, forty lines; each line, approximately eighty black letters. There are also letters on the front cover of each book; those letters neither indicate nor prefigure what the pages inside will say. I am aware that that lack of correspondence once struck men as mysterious. Before summarizing the solution of the mystery (whose discovery, in spite of its tragic consequences, is perhaps the most important event in all history), I wish to recall a few axioms.

First: *The Library has existed* ab æternitate. That truth, whose immediate corollary is the future eternity of the world, no rational mind can doubt. Man, the imperfect librarian, may be the work of chance or of malevolent demiurges; the universe, with its elegant appointments—its bookshelves, its enigmatic books, its indefatigable staircases for the traveler, and its water closets for the seated librarian—can only be the handiwork of a god. In order to grasp the distance that separates the human and the divine, one has only to compare these crude trembling symbols which my fallible hand scrawls on the cover of a book with the organic letters inside—neat, delicate, deep black, and inimitably symmetrical.

Second: *There are twenty-five orthographic symbols.*[1] That discovery enabled mankind, three hundred years ago, to formulate a general theory of the Library and thereby satisfactorily solve the riddle that no conjecture had been able to divine—the formless and chaotic nature of virtually all books. One book, which my father once saw in a hexagon in circuit 15–94, consisted of the letters M C V perversely repeated from the first line to the last. Another (much consulted in this zone) is a mere labyrinth of letters whose penultimate page contains the phrase O *Time thy pyramids.* This much is known: For every rational line or forthright statement there are leagues of senseless cacophony, verbal nonsense, and incoherency. (I know of one semibarbarous zone whose librarians repudiate the "vain and superstitious habit" of trying to find sense in books, equating such a quest with attempting to find meaning in dreams or in the chaotic lines of the palm of one's hand.... They will acknowledge that the inventors of writing imitated the twenty-five natural symbols, but contend that that adoption was fortuitous, coincidental, and that books in themselves have no meaning. That argument, as we shall see, is not entirely fallacious.)

For many years it was believed that those impenetrable books were in ancient or far-distant languages. It is true that the most ancient peoples, the first librarians, employed a language quite different from the one we speak today; it is true that a few miles to the right, our language devolves into dialect and that ninety floors above, it becomes incomprehensible. All of that, I repeat, is true—but four hundred ten pages of unvarying M C V's cannot belong to any language, however dialectal or primitive it may be. Some have suggested that each letter influences the next, and that the value of M C V on page 71, line 3, is not the value of the same series on another line of another page, but that vague thesis has not met with any great acceptance. Others have mentioned the possibility of codes;

---

[1] The original manuscript has neither numbers nor capital letters; punctuation is limited to the comma and the period. Those two marks, the space, and the twenty-two letters of the alphabet are the twenty-five sufficient symbols that our unknown author is referring to. [Ed. note.]

that conjecture has been universally accepted, though not in the sense in which its originators formulated it.

Some five hundred years ago, the chief of one of the upper hexagons[2] came across a book as jumbled as all the others, but containing almost two pages of homogeneous lines. He showed his find to a traveling decipherer, who told him that the lines were written in Portuguese; others said it was Yiddish. Within the century experts had determined what the language actually was: a Samoyed-Lithuanian dialect of Guaraní, with inflections from classical Arabic. The content was also determined: the rudiments of combinatory analysis, illustrated with examples of endlessly repeating variations. Those examples allowed a librarian of genius to discover the fundamental law of the Library. This philosopher observed that all books, however different from one another they might be, consist of identical elements: the space, the period, the comma, and the twenty-two letters of the alphabet. He also posited a fact which all travelers have since confirmed: *In all the Library, there are no two identical books.* From those incontrovertible premises, the librarian deduced that the Library is "total"—perfect, complete, and whole—and that its bookshelves contain all possible combinations of the twenty-two orthographic symbols (a number which, though unimaginably vast, is not infinite)—that is, all that is able to be expressed, in every language. *All*—the detailed history of the future, the autobiographies of the archangels, the faithful catalog of the Library, thousands and thousands of false catalogs, the proof of the falsity of those false catalogs, a proof of the falsity of the *true* catalog, the gnostic gospel of Basilides, the commentary upon that gospel, the commentary on the commentary on that gospel, the true story of your death, the translation of every book into every language, the interpolations of every book into all books, the treatise Bede could have written (but did not) on the mythology of the Saxon people, the lost books of Tacitus.

When it was announced that the Library contained all books, the first reaction was unbounded joy. All men felt themselves the possessors of an intact and secret treasure. There was no personal problem, no

---

[2] In earlier times, there was one man for every three hexagons. Suicide and diseases of the lung have played havoc with that proportion. An unspeakably melancholy memory: I have sometimes traveled for nights on end, down corridors and polished staircases, without coming across a single librarian.

world problem, whose eloquent solution did not exist—somewhere in some hexagon. The universe was justified; the universe suddenly became congruent with the unlimited width and breadth of humankind's hope. At that period there was much talk of The Vindications—books of *apologiæ* and prophecies that would vindicate for all time the actions of every person in the universe and that held wondrous arcana for men's futures. Thousands of greedy individuals abandoned their sweet native hexagons and rushed downstairs, upstairs, spurred by the vain desire to find their Vindication. These pilgrims squabbled in the narrow corridors, muttered dark imprecations, strangled one another on the divine staircases, threw deceiving volumes down ventilation shafts, were themselves hurled to their deaths by men of distant regions. Others went insane.... The Vindications do exist (I have seen two of them, which refer to persons in the future, persons perhaps not imaginary), but those who went in quest of them failed to recall that the chance of a man's finding his own Vindication, or some perfidious version of his own, can be calculated to be zero.

At that same period there was also hope that the fundamental mysteries of mankind—the origin of the Library and of time—might be revealed. In all likelihood those profound mysteries can indeed be explained in words; if the language of the philosophers is not sufficient, then the multiform Library must surely have produced the extraordinary language that is required, together with the words and grammar of that language. For four centuries, men have been scouring the hexagons.... There are official searchers, the "inquisitors." I have seen them about their tasks: they arrive exhausted at some hexagon, they talk about a staircase that nearly killed them—rungs were missing—they speak with the librarian about galleries and staircases, and, once in a while, they take up the nearest book and leaf through it, searching for disgraceful or dishonorable words. Clearly, no one expects to discover anything.

That unbridled hopefulness was succeeded, naturally enough, by a similarly disproportionate depression. The certainty that some bookshelf in some hexagon contained precious books, yet that those precious books were forever out of reach, was almost unbearable. One blasphemous sect proposed that the searches be discontinued and that all men shuffle letters and symbols until those canonical books, through some improbable stroke of chance, had been constructed. The authorities were forced to issue strict orders. The sect disappeared, but in my childhood I have seen old men who for long periods would hide in the latrines with metal disks and a forbidden dice cup, feebly mimicking the divine disorder.

Others, going about it in the opposite way, thought the first thing to do was eliminate all worthless books. They would invade the hexagons, show credentials that were not always false, leaf disgustedly through a volume, and condemn entire walls of books. It is to their hygienic, ascetic rage that we lay the senseless loss of millions of volumes. Their name is execrated today, but those who grieve over the "treasures" destroyed in that frenzy overlook two widely acknowledged facts: One, that the Library is so huge that any reduction by human hands must be infinitesimal. And two, that each book is unique and irreplaceable, but (since the Library is total) there are always several hundred thousand imperfect facsimiles—books that differ by no more than a single letter, or a comma. Despite general opinion, I daresay that the consequences of the depredations committed by the Purifiers have been exaggerated by the horror those same fanatics inspired. They were spurred on by the holy zeal to reach—someday, through unrelenting effort—the books of the Crimson Hexagon—books smaller than natural books, books omnipotent, illustrated, and magical.

We also have knowledge of another superstition from that period: belief in what was termed the Book-Man. On some shelf in some hexagon, it was argued, there must exist a book that is the cipher and perfect compendium *of all other books,* and some librarian must have examined that book; this librarian is analogous to a god. In the language of this zone there are still vestiges of the sect that worshiped that distant librarian. Many have gone in search of Him. For a hundred years, men beat every possible path—and every path in vain. How was one to locate the idolized secret hexagon that sheltered Him? Someone proposed searching by regression: To locate book A, first consult book B, which tells where book A can be found; to locate book B, first consult book C, and so on, to infinity.... It is in ventures such as these that I have squandered and spent my years. I cannot think it unlikely that there is such a total book[3] on some shelf in the universe. I pray to the unknown gods

---

[3] I repeat: In order for a book to exist, it is sufficient that it be *possible.* Only the impossible is excluded. For example, no book is also a staircase, though there are no doubt books that discuss and deny and prove that possibility, and others whose structure corresponds to that of a staircase.

that some man—even a single man, tens of centuries ago—has perused and read that book. If the honor and wisdom and joy of such a reading are not to be my own, then let them be for others. Let heaven exist, though my own place be in hell. Let me be tortured and battered and annihilated, but let there be one instant, one creature, wherein thy enormous Library may find its justification.

Infidels claim that the rule in the Library is not "sense," but "nonsense," and that "rationality" (even humble, pure coherence) is an almost miraculous exception. They speak, I know, of "the feverish Library, whose random volumes constantly threaten to transmogrify into others, so that they affirm all things, deny all things, and confound and confuse all things, like some mad and hallucinating deity." Those words, which not only proclaim disorder but exemplify it as well, prove, as all can see, the infidels' deplorable taste and desperate ignorance. For while the Library contains all verbal structures, all the variations allowed by the twenty-five orthographic symbols, it includes not a single absolute piece of nonsense. It would be pointless to observe that the finest volume of all the many hexagons that I myself administer is titled *Combed Thunder*, while another is titled *The Plaster Cramp*, and another, *Axaxaxas mlö*. Those phrases, at first apparently incoherent, are undoubtedly susceptible to cryptographic or allegorical "reading"; that reading, that justification of the words' order and existence, is itself verbal and, *ex hypothesi*, already contained somewhere in the Library. There is no combination of characters one can make—*dhcmrlchtdj*, for example—that the divine Library has not foreseen and that in one or more of its secret tongues does not hide a terrible significance. There is no syllable one can speak that is not filled with tenderness and terror, that is not, in one of those languages, the mighty name of a god. To speak is to commit tautologies. This pointless, verbose epistle already exists in one of the thirty volumes of the five bookshelves in one of the countless hexagons—as does its refutation. (A number n of the possible languages employ the same vocabulary; in some of them, the *symbol* "library" possesses the correct definition "everlasting, ubiquitous system of hexagonal galleries," while a library—the thing—is a loaf of bread or a pyramid or something else, and the six words that define it themselves have other definitions. You who read me—are you certain you understand my language?)

Methodical composition distracts me from the present condition of humanity. The certainty that everything has already been written annuls us, or renders us phantasmal. I know districts in which the young people prostrate themselves before books and like savages kiss their pages, though they cannot read a letter. Epidemics, heretical discords, pilgrimages that inevitably degenerate into brigandage have decimated the population. I believe I mentioned the suicides, which are more and more frequent every year. I am perhaps misled by old age and fear, but I suspect that the human species—the *only* species—teeters at the verge of extinction, yet that the Library—enlightened, solitary, infinite, perfectly unmoving, armed with precious volumes, pointless, incorruptible, and secret—will endure.

I have just written the word "infinite." I have not included that adjective out of mere rhetorical habit; I hereby state that it is not illogical to think that the world is infinite. Those who believe it to have limits hypothesize that in some remote place or places the corridors and staircases and hexagons may, inconceivably, end—which is absurd. And yet those who picture the world as unlimited forget that the number of possible books is *not*. I will be bold enough to suggest this solution to the ancient problem: *The Library is unlimited but periodic.* If an eternal traveler should journey in any direction, he would find after untold centuries that the same volumes are repeated in the same disorder—which, repeated, becomes order: the Order. My solitude is cheered by that elegant hope.[4]

Mar del Plata, 1941

---

[4] Letizia Alvarez de Toledo has observed that the vast Library is pointless; strictly speaking, all that is required is *a single volume*, of the common size, printed in nine- or ten-point type, that would consist of an infinite number of infinitely thin pages. (In the early seventeenth century, Cavalieri stated that every solid body is the super-position of an infinite number of planes.) Using that silken *vademecum* would not be easy: each apparent page would open into other similar pages; the inconceivable middle page would have no "back."

ONE

Combinatorics

Contemplating Variations of the 23 Letters

There are some, King Gelon, who think that the number of the sand is infinite in multitude; and I mean by the sand not only that which exists about Syracuse and the rest of Sicily but also that which is found in every region whether inhabited or uninhabited. Again there are some who, without regarding it as infinite, yet think that no number has been named which is great enough to exceed its multitude.

—Archimedes, *The Sand Reckoner*

WE BEGIN WITH A PAEAN TO THE MODERN method of denoting numbers, especially the convention of exponential notation, employed first by Descartes in 1637, then extended over the next few decades, primarily by Napier and Newton. (These days, it's commonly also called scientific notation.) In one of his most famous works, Archimedes, a singularly brilliant intellect of the classical world, needed approximately 12 pages (in English translation) to create names of numbers and methods of multiplication to produce an *upper bound*—a maximal estimate, a cap—on the number of grains of sand in the world. By using modern notation, particularly the idea of the exponential, it will take us less than one paragraph to produce an upper bound on the number of grains of sand in the *universe*. Furthermore, in short order these exponential conventions confer the power to accomplish a task that might well have stymied Archimedes: calculating the precise number of distinct books in the Library.

A positive integer *exponent* signifies, "the amount of times some number is multiplied by itself." For example,

$$5^3 = 5 \cdot 5 \cdot 5 \quad \text{and} \quad 2^{4,781} = \underbrace{2 \cdot 2 \cdot \ldots \cdot 2 \cdot 2}_{4,781 \text{ times}}$$

are concise ways to express a "small" number

$$5^3 = 125$$

and a very large number.* There are only two rules regarding the manipulation of exponentials that concern us. The first:

Rule 1: Multiplying numbers written in exponential notation is equivalent to adding the exponents.

For example:

$$5^3 \cdot 5^{14} = (5 \cdot 5 \cdot 5) \cdot \underbrace{(5 \cdot 5 \cdot \ldots \cdot 5 \cdot 5)}_{14 \text{ times}} = \underbrace{(5 \cdot 5 \cdot \ldots \cdot 5 \cdot 5)}_{17 \text{ times}} = 5^{17}.$$

The second rule nicely complements the first.

---

* 16765220490415253625065478163110488777596070684631829708120311409986396665091758868942316900907777384574090574408577882732061772110931659947395687145914975458247961380758354211972797797543235764905722567864684228003984140011308404044321592205678736478798197529921801160919630700034601028770571338599864608382013346981059927132254573497776672384010771401829567908204330728555087268882788756701045666019881331730857746162509298075197595544222542679771939320336753257500121184255659451977833006976704779734418014035299242025994947002632316703732187102015655408002862898537203501628930484732310405790202697134224362089551868316162061097153281907964426167401973307560963972542594814111792976057141050152917573693905714248097057105279956426202806971966214302757930932259278003765598829949253276126891960089208295636389664059681510791937035167989779354104108704854804731802066926964601413195747505371623024014581519128946839050175204929154926102506076582000820459233579973871624581533039027827192594822076477326080909994846009681777529003361408645173508147190013663404830519365501647324846666372695454023369419855605974124635054913613707789078539963199486512143281891270633487234820460978516962245945218404332537360951568826338781616515570835346956655181118415903807293154781056536328031237197140629856224647808737617991705253905525688581305925549191324529576300043914446535610319757557673115929921792891932243531101879093801244638169577763635²

Rule 2: Dividing numbers written in exponential notation is equivalent to subtracting the denominator's exponent from the numerator's.

For example:

$$\frac{2^{4,781}}{2^{14}} = \frac{\overbrace{2 \cdot 2 \cdot \ldots \cdot 2 \cdot 2}^{4,781 \text{ times}}}{\underbrace{2 \cdot 2 \cdot \ldots \cdot 2 \cdot 2}_{14 \text{ times}}} = \frac{\overbrace{2 \cdot 2 \cdot \ldots \cdot 2 \cdot 2}^{14 \text{ times}}}{\underbrace{2 \cdot 2 \cdot \ldots \cdot 2 \cdot 2}_{14 \text{ times}}} \times \frac{\overbrace{2 \cdot 2 \cdot \ldots \cdot 2 \cdot 2}^{4,767 \text{ times}}}{1}$$

$$= \overbrace{2 \cdot 2 \cdot \ldots \cdot 2 \cdot 2}^{4,767 \text{ times}} = 2^{4,767}.$$

The second rule leads to the useful convention of using a *negative* exponent to represent a power in the denominator, for instance,

$$\frac{1}{2^{14}} = 2^{-14}.$$

Thus the previous example may concisely be written

$$\frac{2^{4,781}}{2^{14}} = \left(2^{4,781}\right)\left(2^{-14}\right) = 2^{4,781+(-14)} = 2^{4,767}.$$

It is remarkable that such relatively simple notation can transform relatively complicated tasks, multiplication and division, into the relatively easy and intuitive computations of addition and subtraction.

While pondering previous critical responses to "The Library of Babel," we discovered that a number of people either calculated the number of books or gave some indication of how one might go about it.[1] Our intent in providing the lightning review of exponential notation is to demystify the calculation, and then, more importantly, to give a sense of the enormity of the Library. Then, after the calculation, we tease out a previously overlooked detail from the story and use it to set a new lower bound on the number of books in the Library. (For us, a *lower bound* will be number that says, "We guarantee that *there are at least this many* books in the Library.")

For the purposes of this book, *combinatorics* is the branch of mathematics that counts the number of ways objects can be combined or ordered. Before using combinatorics to calculate the number of the books, let's consider 10 familiar orthographic objects, the symbols we use as representations for digits: 3, 8, 9, 1, 6, 2, 0, 5, 7, 4. We deliberately disordered them to help you see them not as you usually do, as *numbers*, but rather as symbolic representatives of the numbers 0 through 9.

Using these symbols, we'd like to occupy exactly one slot with one symbol, and so we ask: how many distinct ways can we fill one slot? Hopefully, the answer is clear—there are 10 ways to fill one slot with one of the symbols.

1. 0
2. 1
3. 2
4. 3
5. 4
6. 5
7. 6
8. 7
9. 8
10. 9

Now, how many distinct ways are there to fill *two* slots, such that each slot contains one symbol? One complete list of answers, ordered in a familiar way, reads: 00, 01, 02, 03, ..., 97, 98, 99. So we see that there are 100 ways to fill the two slots, given that each slot contains one symbol and that repetition is allowed (enabling such combinations as 00, 11, 22, 33, etc.). Deliberately blurring the distinction between the orthographic symbols and the numbers they represent, we note that there are

$$100 = 10 \cdot 10 = 10^2$$

ways to fill the two slots. If we ask how many distinct ways there are to fill *three* slots, such that repetition is allowed and each slot contains one symbol, we generalize our work from above and produce a complete list that reads: 000, 001, 002, 003, ..., 997, 998, 999. This time, we see that there are 1,000 ways to fill the three slots. Continuing to blur the distinction between the orthographic symbols and the numbers they represent, it follows that there are

$$1,000 = 10 \cdot 10 \cdot 10 = 10^3$$

distinct ways to fill the three slots. By seizing on these ideas, by sensing that a simple pattern has been established and can be used to predict what we couldn't possibly list, we may ask how many distinct ways there are to fill, for example, 36 slots, where each slot contains one of our 10 allowed orthographic symbols and repetition of symbols is allowed. By applying the reasoning we established above, we see that there must be 10^{36} ways; that is, a 1 followed by thirty-six 0s—exactly a billion, billion, billion, billion ways:

$10^{36} = 1,000,000,000,000,000,000,000,000,000,000,000,000.$

Just for a lark, here are the first few and last few slot-fillings of the usual way one would list the fillings.

1. 000000000000000000000000000000000000
2. 000000000000000000000000000000000001
3. 000000000000000000000000000000000002

(Quite a few more!)

(10^{36} − 2). 999999999999999999999999999999999997
(10^{36} − 1). 999999999999999999999999999999999998
10^{36}. 999999999999999999999999999999999999

And that's the end of the list.

In an article in the academic journal *Variaciones Borges*, our ideal reader, Umberto Eco, argues that the exact number of distinct volumes in the Library is irrelevant to both the story and to the reader. To the extent that the numbers of pages, lines, and letters in each book were chosen arbitrarily by Borges, we agree with him. (See the beginning of the chapter "Geometry and Graph Theory" for a quote from Borges regarding this matter.) However, we assert that understanding the combinatorial process that produces the exact number of distinct volumes is both important and relevant to an understanding of the story. So let's apply these ideas to the story and, given the numbers and constraints Borges provides, use them to calculate the number of distinct volumes in the Library.

In "The Library of Babel," Borges writes:

> ... each book contains four hundred ten pages; each page, forty lines; each line, approximately eighty black letters. There are also letters on the front cover of each book; these letters neither indicate nor prefigure what the pages inside will say.

From these lines, we conclude each book consists of $410 \cdot 40 \cdot 80 = 1,312,000$ orthographic symbols; that is, we may consider a book as consisting of 1,312,000 slots to be filled with orthographic symbols. Here a few more excerpts from the next few paragraphs:

> *There are twenty-five orthographic symbols.* That discovery enabled mankind, three hundred years ago, to formulate a general theory of the Library and thereby satisfactorily resolve the riddle that no conjecture had been able to divine—the formless and chaotic nature of virtually all books...
>
> Some five hundred years ago, the chief of one of the upper hexagons came across a book as jumbled as all the others, but containing almost two pages of homogeneous lines. He showed his find to a traveling decipherer, who told him the lines were written in Portuguese; others said it was Yiddish. Within the century experts had determined what the language actually was: a Samoyed-Lithuanian dialect of Guaraní, with inflections from classical Arabic. The content was also determined: the rudiments of combinatory analysis, illustrated with examples of endlessly repeating variations. These examples allowed a librarian of genius to discover the fundamental law of the Library.

This philosopher observed that all books, however different from one another they might be, consist of identical elements: the space, the period, the comma, and the twenty-two letters of the alphabet. He also posited a fact which all travelers have since confirmed: *In all the Library, there are no two identical books.* From those incontrovertible premises, the librarian deduced that the Library is "total"—perfect, complete, and whole—and that its bookshelves contain all possible combinations of the twenty-two orthographic symbols (a number which, though unimaginably vast, is not infinite)—that is, all that is able to be expressed, in every language.

How many distinct books constitute the Library? Each book has 1,312,000 slots, each of which may be filled with 25 orthographic symbols—this is the "variations with unlimited repetition" mentioned above. Again, by employing the ideas outlined above, there are

25 ways to fill one slot,
$25 \cdot 25 = 25^2$ ways to fill two slots,
$25 \cdot 25 \cdot 25 = 25^3$ ways to fill three slots,
and so on,
and so on for 1,312,000 slots.

It follows immediately that there are

$$25^{1,312,000}$$

distinct books in the Library. That's it.

Somehow, it feels all too easy, even anticlimactic, as though instead we should have had to write pages and pages of dense, technical, high-level mathematics, overcoming one complex puzzle after another, before arriving at the answer. But most of the beauty—the elegance—of mathematics is this: applying potent ideas and clean notation to a problem much as the precise taps of a diamond-cutter cleave and husk the dispensable parts of the crystal, ultimately revealing the fire within. (Perhaps we should have ended the calculation by writing "That's it!" instead of "That's it.")

Our new twist on these calculations involves what Hurley translates as the "letters on the front cover of each book." For the sake of precision, we note that the Spanish reads "el dorso de cada libro," which translates literally as "the back of the book." Idiomatically and bibliographically, however, the sense of this phrase is that the letters are on the *spine* of the Library's books. As such, the interpretation we use for the rest of this book is that the letters are on the spine.

Now, the number $25^{1,312,000}$ we calculated above doesn't account for these spinal letters. It strikes us as likely that, within the imaginary universe of the Library, a book with the letters *The Plaster Cramp* written on the spine, whose 1,312,000 slots are filled by the repeated sequence of orthographic symbols MCV, should be considered as a book distinct from one with the exact same pages which is instead imprinted with the letters *Axaxaxas Mlö* on the spine.[2] Scanning through the original Spanish version, "La biblioteca de Babel," we find a book described with the 19 orthographic symbols *El calambre de yeso* on its spine. This means that there are a minimum of 19 slots to fill on each spine, and accounting for these variations with repetition expands the Library by a factor of *at least*

$$25^{19} = 363,797,880,709,171,295,166,015,625.$$

We write this number out explicitly to re-echo the vastness of the numbers woven through the Library. Simply adding 19 orthographic symbols on the spine magnifies the Library more than 300 septillion times. For comparison, this number is roughly the number of microscopic plant cells comprising a grove of 364 oak trees.[3] So if the Library of $25^{1,312,000}$ books is considered as *one* imperceptible plant cell, accounting for differing symbols on the spine multiplies the Library into a *grove* of 364 giant oak trees.

However, since we cannot be sure of either the maximum number of symbols on the spine of each book or of Borges' intent, we restrict ourselves to $25^{1,312,000}$ books. This number, so easy to write, is, in a powerful sense, utterly unimaginable. To see that we can't see it, let's begin by converting this number to a *power of 10*, which puts it in a more familiar context.

$25^{1,312,000}$ is just a little bit larger than $10^{1,834,097}$;

which is, of course, a 1 followed by one million, eight hundred thirty-four thousand, and ninety-seven 0s. We accomplish this conversion to a power

FIGURE 3. Our universe, represented as a cube.

of 10 notation using the *logarithmic* function and discuss the mechanics in the Math Aftermath portion of this chapter.

Could our universe possibly contain the Library? Current research approximates the size of the universe as being about 1.5×10^{26} meters across. Let's simplify calculations and create an upper bound to the universe by overestimating its size and supposing that our universe is shaped like a cube, each side of which measures 10^{27} meters (figure 3).

So, we'll say that our cubic universe consists of approximately $10^{27} \cdot 10^{27} \cdot 10^{27} = 10^{81}$ cubic meters. If we assume that we may fit $1{,}000 = 10^3$ Library books in a cubic meter—and this is an exceedingly generous assumption—then our universe, if it consisted of *nothing* except books, would contain

$$10^{81} \cdot 10^3 = 10^{84} \text{ books.}$$

This doesn't make the slightest dent in the Library; it would take

$$\frac{10^{1{,}834{,}097}}{10^{84}} = 10^{1{,}834{,}097 + (-84)} = 10^{1{,}834{,}013}$$

universes the size of ours to hold just the books of the Library. What if the books were each as small as a grain of sand?

Using a ruler shows that an average grain of sand is approximately one millimeter across. If we assume a cubical shape combined with a perfect packing, then we could fit approximately

$$10^3 \cdot 10^3 \cdot 10^3 = 10^9 = 1{,}000{,}000{,}000 = \text{one billion}$$

grain-of-sand books in a cubic meter. Multiplying by the size of the universe, we find that the universe holds only

$$10^{81} \cdot 10^9 = 10^{90} \text{ such books.}$$

That is, if the universe consisted of nothing but sand, it would hold at most about 10^{90} grains of sand. As we promised at the beginning of the chapter, using exponential notation allows us to estimate the number of grains of sand considerably faster than Archimedes.

Once again, though, this hardly impacts the Library's collection. As a final illustration of this point, suppose that each book is shrunk to the size of a proton; that is, shrunk to about 10^{-15} meters across. Given that each book is 10^{-15} meters across, we could pack 10^{15} of them in a very narrow one-meter-long strip. Thus, packing a cubic meter with proton-sized books yields

$$10^{15} \cdot 10^{15} \cdot 10^{15} = 10^{45} \text{ books.}$$

Our universe holds merely

$$10^{81} \cdot 10^{45} = 10^{126} \text{ of these subatomic books.}$$

Let's adopt one more viewpoint in our efforts to conceptualize the enormity and complexity of the Library. Perhaps the simplest books to imagine, of which there are exactly 25, are those that consist of nothing except one symbol, repeated for the entire book. For example, one such book would consist of its 1,312,000 slots filled by the letter g.[4] The first two lines of that book would read

ggggggggggggggggggggggggggggggggggggggggggggggggggggggggggggggggggg
ggggggggggggggggggggggggggggggggggggggggggggggggggggggggggggggggggg

and so on for another 38 lines on the first page, followed by 40 more lines on each of the remaining 409 pages: a veritable rhapsody in g.

Now allow a slight variation. The next set of books we consider are those that consist of entirely the orthographic symbol g except for one h. That is, exactly 1,311,999 slots will be filled with the letter g, while exactly one slot will contain the letter h. One such book will begin

ggggggggggggggggggggggggggggggggggggggggggggggggggggggggggggggggggg
ggggggggggggggggggggggggggggghgggggggggggggggggggggggggggggggggggg

and, as above, all of the rest of the symbols in the book are the letter g.

How many books like this are there? Well, there are exactly 1,312,000 different slots that the single h can occupy, and every other slot must be filled with a g. Thus, there are exactly 1,312,000 such books.

Now, we allow ourselves to imagine a book that consists of 1,311,998 slots filled with the symbol g and two slots—not necessarily adjacent—filled with an h. There are precisely

$$\frac{(1,312,000) \cdot (1,311,999)}{2} = 860,671,344,000$$

such books. (At the end of this chapter, see the second Math Aftermath, "An Example of the Ars Combinatoria," for an explanation of this and the next two formulas.) Put into human terms, assuming the world population is currently somewhat less than seven billion people, this translates to every one of us enjoying a personal library of about 123 of these books.

If we next consider books that, excepting three instances of the letter h, are all g, all the time, we perform a similar calculation to find that there are exactly

$$\frac{(1,312,000) \cdot (1,311,999) \cdot (1,311,998)}{6} = 376,399,693,995,104,000$$

such books. This number, although perhaps not appearing much larger than the preceding one, expands these monotonous libraries to about 53 million distinct books for each person currently alive.

Pursuing this notion to its conclusion, by considering the number of books consisting of a mere 16 occurrences of the letter h in an otherwise uniform desert of the letter g, we find there are

3,683,681,259,485,362,310,918,865,543,989,208,654,728,931,149, 486,911,733,618,072,454,576,141,229,488,660,718,000

distinct books—about 3.7×10^{84} books—more than enough to fill three cubic universes. These books, droning wearily of g with a little respite provided only by the scant 16 instances of h, are not typographical phantasmagoria to inflame the imagination or addle the senses, and yet if they were all collected into a subsection of the Library, they would occupy a space greater than three times our known universe.

Finally, it would be a tedious, uninspired, but straightforward calculation to determine how big the Library needs be to hold the books in the hexagonal configurations described by Borges. Given the work we've just done, it should be clear that however the Library is constructed, any sort of ambulatory circumnavigation would be utterly impossible for a human being: a vigorous, long-lived librarian who managed to walk a little over 60 miles—about 100 kilometers—every day for 100 years would cover somewhat less distance than light travels in twelve *seconds*. To cross our universe, which is incomprehensibly dwarfed by the Library, light would need to travel for at least 15 billion *years*.

The number of books in the Library, although easily notated, is unimaginable.

Math Aftermath I: The Logos of Logarithms

> *There are those who dance to the rhythm that is played to them, those who only dance to their own rhythm, and those who don't dance at all.*
> —José Bergamín, *The Rocket and the Star*

This Aftermath is included for two purposes, one explanatory and one hortatory. The expository side is to provide a basis for those who wish to understand the details of how certain approximations and calculations are made in this chapter, as well as the chapter "Real Analysis." The public relations portion is to reconceive of the logarithm as a function imbued with a friendly collection of useful, easily manipulated properties.

For the purposes of this book, we'll say that a *function* is a rule such that for each legitimate number the rule is applied to, it returns back exactly one number. The output number might be the same or different from the input number; however, the important thing is that given a specific input number, the output number for it is always the same. (There are many interesting generalizations of this idea, including that of studying spaces whose elements are themselves functions.) One of the functions most misunderstood and maligned by generations of students is that of the logarithm.

The logarithm (base 10) is typically notated *log*; frequently it is written $log(x)$ to emphasize it is a function: given an input of one number, x, it outputs another number, $log(x)$. The modern notation is quite evocative:

$$x \to log(x).$$

We could, at this juncture, include a graph of the logarithmic function; after all, a picture is useful for nurturing our visual awareness. However, we deliberately exclude such an illustration to hammer home a point: the logarithm, as it turns out, is a function that may be defined by a number of truly remarkable properties. Since really we only need to use one of the properties, let's jump right in: if x is any positive number, and n is any number, then

$$log(x^n) = n \cdot log(x).$$

That is, the logarithm, remarkably, "lowers" the exponential, thereby reducing it to a much more familiar operation—multiplication. There are many marvelous implications of this property, but for our purposes, the property alone will give us what we need.

Earlier in the chapter, using exponential notation, we found that there are $25^{1,312,000}$ distinct volumes in the Library. We'd like to contextualize the number of books by putting that number into a somewhat more familiar form. We choose to convert it to the power of 10 notation, 10^n, because we may think of that as a single 1 followed by n 0s. Therefore, we set up the following equation and endeavor to solve it for n.

$$25^{1,312,000} = 10^n$$

When we solve this equation for n, we thus gain a greater intuition for the number of books in the Library.

Here's the key point: even though $25^{1,312,000}$ and 10^n are written differently and look different, if we choose some n such that the two numbers are equal, *then they are, in fact, the same number.* Since they are the same number, by the definition of a function, it must the case that using both representations of the number as inputs to the function entails that both of the outputs must continue to be equal to each other. So we apply the logarithm to both sides of the equation and get

$$log\left(25^{1,312,000}\right) = log\left(10^n\right).$$

Now, the remarkable property of the logarithm "brings the exponential down" and gives

$$1{,}312{,}000 \cdot \log(25) = n \cdot \log(10).$$

(In fact, things are even better than they appear, for $\log(10)$ is equal to 1, but that need not concern us here.) Divide both sides by $\log(10)$ to solve for n:

$$1{,}312{,}000 \cdot \frac{\log(25)}{\log(10)} = n.$$

By using a calculator, a computer, or even Henry Brigg's log tables from 1617, we find

$$n \approx 1{,}834{,}097.$$

Therefore,

$$25^{1{,}312{,}000} \approx 10^{1{,}834{,}097}.$$

Math Aftermath II: An Example of the Ars Combinatoria

Drawing is a struggle between nature and the artist, in which the better the artist understands the intentions of nature, the more easily he will triumph over it. For him it is not a question of copying, but of interpreting in a simpler and more luminous language.
—Charles Baudelaire, *The Salon of 1846*, VII. "On the Ideal and the Model"

In the final analysis, a drawing simply is no longer a drawing, no matter how self-sufficient its execution may be. It is a symbol, and the more profoundly the imaginary lines of projection meet higher dimensions, the better.
—Paul Klee, *The Diaries of Paul Klee 1898–1918*, no. 681, entry for July 1905

Here, we endeavor to explain the origins of the (possibly) mysterious formulas appearing earlier in the chapter. The first one arises in the context of trying to determine the number of distinct books in the Library consisting of 1,311,998 occurrences of the letter g and two instances of the letter h.

We abstract the books and hexagons away from the problem by noting that what we are really interested in can be characterized as the question "How many distinct ways exist to pick two objects from 1,312,000?" The two objects, of course, correspond to the two slots that we will fill with the letter h. So, the number of distinct ways to choose two objects from 1,312,000 corresponds precisely to the number of distinct books under discussion.

As it turns out, for millennia combinatorialists have known a formula for this and related questions; in the most general terms, the number of different ways to choose a subset of k objects from a set of n objects is equal to

$$\frac{\overbrace{(n) \cdot (n-1) \cdot (n-2) \cdot \cdots \cdot (n-(k-1))}^{k \text{ terms}}}{k!}.$$

One way to uncover the derivation of this formula is to break the analysis into two parts, first explaining the terms appearing in the numerator, and then understanding the term in the denominator. (Joe Roberts, the professor who introduced me to combinatorial analysis, helpfully said "attic" and "basement" instead of "numerator" and "denominator.")

We wish to choose one object from n distinct objects. Thus, we have n choices for our first object and then we are left with $n-1$ objects. So, when we choose the second object, we have $n-1$ distinct objects to choose from. This means that choosing *two* objects is tantamount to having $(n) \cdot (n-1)$ ways to pick them: n ways to choose the first object multiplied by $(n-1)$ ways to choose the second.

If we pick a third object, we are choosing from $(n-2)$ distinct objects, and so the numerator grows accordingly. Notice that when we pick a fourth object, we choose from $(n-3)$ distinct objects; thus, extending the developing pattern, when we pick the kth object, we are selecting it from the remaining $(n-(k-1))$ distinct objects. Multiplying, in succession, all of the choices yields the numerator (attic):

$$\overbrace{(n) \cdot (n-1) \cdot (n-2) \cdot \cdots \cdot (n-(k-1))}^{k \text{ terms}}.$$

At this juncture, it's reasonable to wonder why there needs to be a denominator (basement). Why can't we simply stop at the numerator, or, put another way, what is wrong with what we've derived? The answer is devilishly simple: there are a number of different ways to pick the exact same subset of size k, and we don't care in what order the k objects are chosen. We just want to know which are the chosen ones.

Let's illustrate this with an easy example. We have a *set*, a collection, of three distinct objects, {A, B, C}. Let's choose all *subsets* consisting of two distinct objects:

{A, B}   {B, A}
{A, C}   {C, A}
{B, C}   {C, B}

If the order in which the objects are picked is important, then we have a complete list of all subsets of size two. However, if order is unimportant, then {A, B} and {B, A} are both names for the same subset. Really, then, we would be happy with, say, this list.

{A, B}
{A, C}
{C, B}

Since all we care about is the number of ways to choose two things, and we don't care about the order, we need to divide out by the number of repetitions, which in this case, is two. We thus arrive at the complete formula for this example,

$$\frac{\text{(first choice)} \cdot \text{(second choice)}}{\text{repetitions}} = \frac{3 \cdot 2}{2} = 3.$$

Another way to think about repetitions is as the number of distinct orderings of, for instance, a set of k objects. There is a critical difference between this and the exponential work from earlier in the chapter; when calculating the number of books, we allowed an orthographic symbol to be used over and over and over again—possibly 1,312,000 times. Or, conversely, a symbol didn't need to appear at all. In an ordering, every object needs to appear exactly once. In the beginning of the chapter "More Combinatorics," we show how to calculate the number of distinct orderings of a set of k objects: it's a product of k integers, notated $k!$ and pronounced "k factorial." For now, suffice it to note that

$$k! = k \cdot (k-1) \cdot (k-2) \cdot (k-3) \cdot \cdots \cdot 4 \cdot 3 \cdot 2 \cdot 1.$$

This explains the denominator of the formula: we divide out by all the repetitions given by all the different orderings of the k chosen objects and thus achieve a masterpiece of the ars combinatoria:

$$\frac{(\text{first choice}) \cdot (\text{second choice}) \cdot \cdots \cdot \left(k^{\text{th}} \text{ choice}\right)}{\text{repetitions}}$$

$$= \frac{(n) \cdot (n-1) \cdot \cdots \cdot (n-(k-1))}{k!}.$$

Applying this formula to the situation of choosing subsets of size 16 from a set of size 1,312,000 yields the expression

$$\frac{(1{,}312{,}000) \cdot (1{,}312{,}000 - 1) \cdot \cdots \cdot (1{,}312{,}000 - (16-1))}{16!}$$

$$= \frac{(1{,}312{,}000) \cdot (1{,}311{,}999) \cdot \cdots \cdot (1{,}311{,}986) \cdot (1{,}311{,}985)}{16 \cdot 15 \cdot 14 \cdot 13 \cdot 12 \cdot 11 \cdot 10 \cdot 9 \cdot 8 \cdot 7 \cdot 6 \cdot 5 \cdot 4 \cdot 3 \cdot 2 \cdot 1}$$

$$= \frac{(1{,}312{,}000) \cdot (1{,}311{,}999) \cdot \cdots \cdot (1{,}311{,}986) \cdot (1{,}311{,}985)}{20{,}922{,}789{,}888{,}000}$$

$$= 3{,}683{,}681{,}259{,}485{,}362{,}310{,}918{,}865{,}543{,}989{,}208{,}654{,}728{,}$$
$$931{,}149{,}486{,}911{,}733{,}618{,}072{,}454{,}576{,}141{,}229{,}488{,}660{,}$$
$$718{,}000$$

$$\approx 3.7 \times 10^{84}.$$

TWO

Information Theory
Cataloging the Collection

> It is a very sad thing that nowadays there is so little useless information.
> —Oscar Wilde, "A Few Maxims for the Instruction of the Over-Educated"

INFORMATION THEORY IS ONE OF THE YOUNGEST fields in mathematics, essentially born in 1948 when Claude Shannon published "A Mathematical Theory of Communication." As a discipline, it is still unfolding, still crystallizing into a way to analyze and interpret the world. For the purposes of this book, we'll say that information theory is the study of the compression and communication of complex information. We consider each book in the Library to be a complex piece of information, and our inquiry takes the form of investigating how a catalogue of the Library might encode information about the content and location of books. Since the story was written while Borges was tasked with cataloguing the collection of the Miguel Cané Municipal Library, questions of this nature may have taken on rich significance for him.

Typically, a library catalogue card, either physical or virtual, contains two distinct kinds of information. The first sort uniquely specifies a book in such a way that a reader with partial or incomplete information still might identify the book: a title, author, edition, publisher, city of publication, year of publication, and short description of the contents generally appear on a card and prove sufficient. An ISBN also uniquely specifies a book, but probably isn't much use in finding a book if we remember only a few digits of the number.

The second type of information uniquely specifies a location in the library, although additional knowledge is usually required. For example, under most systems of cataloguing currently in use, the call numbers, in addition to uniquely specifying a book, include an abundance of letters and digits, often interspersed with decimal points. If one does not know, say, where the PQ books are shelved, the information is degraded. Even if the books were arranged alphabetically by author or title, for a large collection we'd still need to know in what general region to begin our search. By analogy, many dictionaries have thumbnail indentations which enable readers quickly to find a section of words beginning with one or several letters. Both of these categories of information are problematic for the Library of Babel.

A form a catalogue might take in principle is: Book (identifiers), Hexagon (location), Shelf (only 20 per hexagon), Position on Shelf (only 32 books per shelf). Perhaps surprisingly, *self-referentiality* is not a problem. A volume of the catalogue, say the tenth, residing in Hexagon 39, Shelf 20, Position 14, could well be marked on the spine "Catalogue Volume Ten," and correctly describe itself as the tenth volume of the catalogue and specify its location in Position 14, on Shelf 20, in Hexagon 39: there is no paradox. However, beginning with the obvious, here are some of the difficulties that arise.

Clearly, the Library holds far too many books to be listed in one volume; any catalogue would necessarily consist of a vast number of volumes, which, perversely, are apt to be scattered throughout the Library. Indeed, reminiscent of the approach of another of Borges' stories, "The Approach to Al-Mu'tasim," and of the lines in "The Library of Babel,"

> To locate book A, first consult book B, which tells where book A can be found; to locate book B, first consult book C, and so on, to infinity....

an immortal librarian trying to track down a specific book likely has a better chance by making an orderly search of the entire library, rather than finding a true catalogue entry for the book. Every plausible entry from any plausible candidate catalogue volume would have to be tracked down, including regressive scavenger hunts. An immortal librarian would spend a lot of time traversing the Library, ping-ponging back and forth between different books purporting to be volumes of a true catalogue.

After revealing the nature of the Library, the librarian notes that contained in the Library are "the faithful catalog of the Library, thousands and thousands of false catalogues, the proof of the falsity of those false catalogs, a proof of the falsity of the true catalogue..." This, then, is the second problem of any catalogue: the only way to verify its faithfulness would be to look up each book. Furthermore, the likelihood of any book being located within a distance walkable within the life span of a mortal librarian is, to all intents and purposes, *zero*. Sadly, even if we were fortunate enough to possess a true catalogue entry for our Vindication, presumably our Vindication would merely give details of the death we encountered while spending our life walking in a fruitless attempt to obtain the Vindication. (Recall in "The Library of Babel," Borges describes Vindications as "books of apology and prophecy which vindicated for all time the acts of every man in the universe and retained prodigious arcana for his future.")

Let's consider the first category of information found on library cards, that which uniquely specifies the book. *Authorship is moot.* One might argue that the God(s), or the Builder(s) of the Library, is (are) the author(s) of any book. One might also make a claim that the author is an algorithm embodied in a very short computer program which would, given time and resources, generate all possible variations of 25 orthographic symbols in strings of length 1,312,000. One could make the Borgesian argument that One Man is the author of *all* books.

For that matter, the writer Pierre Menard, a quixotic character in Borges' story "The Don Quixote of Pierre Menard," may as well be credited with authorship of all the books in the Library.

Certainly there are many, many books whose first page resembles the one in figure 4. How many such books? Specifying one page means that 80 symbols for each of 40 lines are "frozen." This means that out of the 1,312,000 symbols of a book, the first 3,200 are taken, leaving 1,308,800 spaces to fill. By the work of the "Combinatorics" chapter, there are thus precisely $25^{1,308,800}$ books with a first page exactly the same as the depicted title page. (Using logarithms as in the first Math Aftermath, this number is seen to be approximately $10^{1,829,623}$ books.) Viewed from a

THE LIBRARY OF BABEL

BY JORGE LUIS BORGES

A.K.A. CONCISE ALGORITHM
A.K.A. ONE MAN
A.K.A. PIERRE MENARD

LIBRARY OF BABEL PRESS

HEXAGONAL UNIVERSE

FIGURE 4. The first page of many, many books in the Library.

complementary angle, there are $25^{3,200}$ possible first pages, and although significantly smaller than the numbers we've been contemplating, it is yet another enormous number. The chance of randomly selecting a book with this particular first page is "only" 1 in $25^{3,200}$, approximately $10^{4,474}$, which means, essentially, that it will never happen. For comparison's sake, the chance of a single ticket winning a lottery is better than 1 in $100,000,000 = 10^8$. So finding such a book is equivalent to winning the lottery more than 559 times in a row. (In the equation below, each factor of 10^8 signifies winning the lottery once.)

$$\underbrace{10^8 \cdot 10^8 \cdot 10^8 \cdots\cdots 10^8 \cdot 10^8}_{559 \text{ terms}} = 10^{8+8+8+\cdots+8+8} = 10^{559 \cdot 8} = 10^{4{,}472}$$

As a source of useful information for a catalogue entry, a title on the spine of a book, such as *The Plaster Cramp*, is similarly moot, for there must still be something like

$$\frac{\text{Number of distinct books}}{\text{Number of distinct spines}} = \frac{25^{1{,}312{,}000}}{25^{19}} = 25^{1{,}312{,}000-19} = 25^{1{,}311{,}981}$$

distinct books with the exact same orthographic symbols on the spine.

Edition, publisher, city of publication, year of publication—all are meaningless in this Library. The one sort of information we mentioned that may possibly prove useful is that of a short description of the contents of the book. We'll take "short" to mean "half-page or less." It's much more difficult to say what we mean by "description." We'll take it to mean "something that significantly narrows the possible contents of the book." For example, "The book is utter gibberish, completely random nonsense," doesn't significantly narrow the possible contents of the book. (We are aware that this definition is problematic.)

Any book published in the last 500 years likely has a short, reasonably limiting description. A book whose contents consist of the letters MCV repeated over and over evidently has a short description. A book whose entire contents are similar to the 80-symbol line

unmenneo .ernreiuhr. naper, utuytgn or fgioe, no,e,dn .roih senoi.,erg n cprih npp

almost certainly doesn't have a short description. Or does it? A fascinating area of study in the field of information theory concerns the difficulty of deciding whether or not a line such as the one above has some sort of algorithmic description that is shorter than the line itself. Borges seems to have an intimation of this when he writes "There is no combination of characters one can make—*dhcmrlchtdj*, for example—that the divine Library has not foreseen and that in one or more of its secret tongues does not hide a terrible significance." Perhaps "*hr,ns llrteee*" is a more concise description of the line, or perhaps a succinct translation into English is "Call me Ishmael."

It does no good to excerpt a passage as a short description; titanic numbers of books in the Library will contain the same passage. In an important sense, then, for all languages currently known by human beings, for the cataclysmic majority of books in the Library, *the only possible description of the book is the book itself.* This, in turn, leads to a lovely, inescapable, unimagined conclusion:

The Library is its own catalogue.

Let's restrict the investigation to a slightly more agreeable collection of books: all those whose entire contents cohere and are recognizably in English, and whose first page contains precisely a short title and a half-page description, both of which accurately reflect the contents. Any rule of selection will have problems. Some associated with this one are: What does it mean to "cohere"? Would a collection of essays on different topics constitute a coherent work? Would sections of James Joyce's infamous novel *Finnegans Wake* register as "recognizably English"? What if the book contains a non-English word, such as "ficciones"? What if the title, as in the case of *Ulysses*, is more allusive than descriptive? Can any description "accurately reflect" the contents of a book? Regretfully, we'll ignore these and other legitimate, interesting concerns.

For example, suppose the first page of a volume of the Library began with the following description, modified slightly from the back cover of the 2002 Routledge Press edition of Wittgenstein's *Tractatus Logico-Philosophicus.*

> Tractatus Logico-Philosophicus
> by Ludwig Wittgenstein
>
> Perhaps the most important work of
> philosophy written in the twentieth
> century, Tractatus Logico-Philosophicus
> was the only philosophical work that
> Ludwig Wittgenstein published during his

lifetime. Written in short, carefully numbered paragraphs of extreme brilliance, it captured the imagination of a generation of philosophers. For Wittgenstein, logic was something we use to conquer a reality which is in itself both elusive and unobtainable. He famously summarized the book in the following words, "What can be said at all can be said clearly, and what we cannot talk about we must pass over in silence."

If next came the precise contents of the book, including Bertrand Russell's introduction, followed by the appropriate number of pages consisting of nothing but blanks, then that Library volume would be included in the collection. We are also willing to include books longer than 410 pages, so long as the title page includes reference to an appropriate volume number. This allows, among other things, for the inclusion of this Catalogue of Books in English into the putative catalogue we are trying to define, which we may as well call *Books in English*.

This amenable collection of books is designed to enable *Books in English* to include a title and short accurate description of the contents. This nearly accomplishes the first half of the task of a catalogue; although the books aren't uniquely specified, the scope of possibility is greatly constricted. However, the other half of a catalogue, that of specifying a location, is also fraught with difficulties.

First, and most strongly emphasized by Borges, is the apparent lack of organization in the distribution of books. It is possible that there is an overarching pattern, but even if there is, it would be impossible to deduce it from local information. The librarian's "elegant hope" that the Library is (truly) infinite and periodic would provide a godlike observer with a kind of an order for each book; every particular book would have an infinite number of exact copies—unimaginably distant from each other—and these infinite copies would constitute a set of regularly spaced three-dimensional lattice points. But this pattern does not serve our needs.

Finite or infinite, the problem of identifying individual hexagons of the Library is insurmountable. If the Library is a 3-sphere or any of the other spaces described in the chapter "Topology and Cosmology," the number of hexagons is finite. However, since each hexagon holds 640 books, which is approximately $25^{2.007}$ books, more than $25^{1,311,997}$ (approximately $10^{1,834,095}$) hexagons are required to hold all the Library's books. This means that if one were to attempt to write out a number for each hexagon in our familiar base-10 notation, it would take 1,834,095 *digits*. Now each book in the Library has exactly 1,312,000 slots to fill, and, moreover, the orthographic symbols contain no (recognizable) digits. Writing a number out in words usually uses many more precious slots; for example,

> [one million, eight hundred thirty four thousand, and ninety five] versus 1,834,095.

The bracketed expression takes 63 spaces, while the second needs only nine. For almost every hexagon in the Library, a volume of a hypothetical *Books in English* catalogue could not actually contain the corresponding hexagon number where a book is shelved.

Trying to circumvent this problem, one may observe that many numbers have shorter expressions, such as $2^{4,781}$, and legitimately wonder if *every* integer might have a remarkably condensed form. An insuperable problem is that there are many such condensed expressions, including the one above, that need a computer to calculate. More disturbing, though, is an example of a condensed verbal description of a "small" number—only 100 digits—that even we, using networked supercomputers, would be unable to find:

> The *median* of the prime numbers expressible in 100 digits.

Thus, even if the catalogue entry for the *Tractatus Logico-Philosophicus* listed the location as

> Hexagon: the median of the prime numbers expressible in one hundred digits.
> Shelf: four.
> Position: seventeen.

the information is as useless to us as it is to a librarian. (See the Math Aftermath "Numb and Number (Theory)" for more discussion about prime numbers and, in particular, why we are unable to determine the median of the prime numbers expressible in 100 digits.)

Usually, outside of computer science, we use *base* 10 to represent the positive integers, meaning we use the 10 symbols {0, 1, 2, 3, 4, 5, 6, 7, 8, 9} to label numbers. In these circumstances, though, one might try using a higher base than 10 for the integers, such as base 25, to number the hexagons. There are two problems associated with this: first, it would still take all but two slots of a book to list a hexagon number, which suffices to invalidate the usefulness. Second, since each book contains only 25 orthographic symbols, each such symbol would have to stand for a digit. So, if one were to write out the hexagon number in base-25 digits, it would usually look like complete gibberish. (In fact, it also leads to an unpleasant, yet valid, interpretation of the Library: it is the complete listing of all base-25 numbers comprised of exactly 1,312,000 digits.) At any rate, such a book would not be "recognizably English"; thus it would not itself be listed in *Books in English*.

What if, like Ireneo Funes, from Borges' celebrated short story "Funes the Memorious," we resolved to work in base 24,000? It would do no good: in the story, for each number up to 24,000 Funes created his own signifier, for example, names such as *Brimstone*, *Clubs*, and *The Whale*. In the Library, we are stuck with 25 orthographic symbols. Instead of combining 10 digits in various ways to fill five places to make a number between 1 and 24,000, we would need to combine the 25 symbols in a minimum of four places to distinguish 24,000 separate numbers, because

$$25^4 = 390,625$$

while

$$25^3 = 15,625$$

which doesn't provide enough distinct signifiers to take us up to base 24,000. Anyway, not only wouldn't this convention save much space, it also leads back to the previous dilemma: writing out the names of the numbers will result in waterfalls of gibberish.

Finally, a potential catalogue entry might take a different tack. It might give coordinates, such as, "Go up ninety-seven floors, move diagonally left four thousand hexagons, and then move diagonally right another two hundred and twenty." Although this might, at first blush, seem appealing, the same sorts of problems arise, for most hexagons are unimaginably far away. The example provided above works simply because the numbers involved—97, 4,000, and 220—are so miniscule, so accessible. The Library is neither.

The Library is its own catalogue. Any other catalogue is unthinkable.

Math Aftermath: Numb and Number (Theory)

> *A metaphysician is one who, when you remark that twice two makes four, demands to know what you mean by twice, what by two, what by makes, and what by four. For asking such questions metaphysicians are supported in oriental luxury in the universities, and respected as educated and intelligent men.*
>
> —H. L. Mencken, *A Mencken Chrestomathy*

Below are two outgrowths from the sprawling yet spare field of number theory; together they form a pair of relatively straightforward mathematical confections. Both revolve around using prime numbers decisively to reach interesting conclusions.

Consider the $25^{1,312,000}$ distinct volumes in the Library: a simple rethinking of this number will produce a result surely unimagined by Borges. Now, as we all know, the number 25 *factors* into 5·5, so

$$25^{1,312,000} = \underbrace{25 \cdot 25 \cdot 25 \cdots \cdot 25}_{1,312,000 \text{ terms}} = \underbrace{(5 \cdot 5) \cdot (5 \cdot 5) \cdot (5 \cdot 5) \cdots \cdot (5 \cdot 5)}_{2 \cdot (1,312,000) = 2,624,000 \text{ terms}}$$

$$= 5^{2,624,000}.$$

A *prime number* is a positive integer greater than one that is divisible only by itself and by one. The unique factorization theorem, proved by Euclid in *The Elements*, says that every positive integer is decomposable into exactly one *product* of primes, each of which is raised to a power greater than or equal to one. For example, we all know that 100 = 10 · 10, and it's also true that 100 = 4 · 25. So, what is 100 equal to, 10 · 10 or 4 · 25? Of course you're laughing at us, because 100 is obviously equal to both products. Neither of these answers, though, is written exactly as a product of primes, in which each prime is raised to a power greater than or equal to one. Based on the two factorizations—10 · 10 and 4 · 25—it's easy to see that

$$100 = 10 \cdot 10 = (2 \cdot 5) \cdot (2 \cdot 5) = (2 \cdot 2) \cdot (5 \cdot 5) = (2^2) \cdot (5^2)$$

and

$$100 = 4 \cdot 25 = (2 \cdot 2) \cdot (5 \cdot 5) = (2^2) \cdot (5^2).$$

Because 100 is so familiar, it's probably not surprising to you that both of the initial factorizations lead to the unique one. And perhaps it is equally intuitive that no matter how large an integer we begin with, no matter how we might try, there will be only one way to factor it into powers of primes. Still, it's nice to know that Euclid showed that it must always be true.

By the work above, $5^{2,624,000}$ is a unique factorization of $25^{1,312,000}$ into primes, each raised to a power greater than or equal to one. In this case, plainly the number of distinct books uniquely decomposes to one prime (5) raised to a power greater than one (2,624,000). It follows that the only numbers that can divide $25^{1,312,000}$ are powers of five. Now, as is easily inferred from the story, each hexagon in the Library contains 640 books. The number 640 uniquely factors into $2^7 \cdot 5$, and so the number 640 does not divide $25^{1,312,000}$, for

$$\frac{25^{1,312,000}}{640} = \frac{5^{2,624,000}}{640} = \frac{5 \cdot 5^{2,623,999}}{5 \cdot 2^7} = \frac{5^{2,623,999}}{2^7},$$

and none of the seven 2s in the denominator may divide any of the millions of 5s in the numerator. This means that the books do not exactly fill out all the hexagons, which entails that either the Library is *not complete* (!!!), or that there is a special hexagon that is not full, or that at least one hexagon is differently configured, or that at least one hexagon contains exact copies of other books in the Library. We can't imagine that Borges considered this—or would have cared—when he assigned numbers to the quantity of shelves on a wall or the number of books per shelf in the Library.

Also, it may seem easy to juggle and tweak the numbers of shelves and books to make each hexagon hold, say, $625 = 5^4$ books. After all, as written in the story, each hexagon holds 640 books, and 625 is very close to 640. But this is an opportunity to admire the power of Euclid's unique factorization theorem: if each of the four non-doorway walls has the same number of shelves, and if each shelf holds the same number of books, then each hexagon must hold

(4 walls) × (m shelves per wall) × (n books per shelf) = $4mn$ books.

The prime factors $2^2 = 4$ will always be there; neither adjusting the number of shelves per wall, nor the tally of books per shelf will budge those 2s, which means that $4mn$ can never cleanly divide $25^{1,312,000}$.

How, then, might we arrange matters so that the total number of distinct volumes may be evenly distributed throughout the hexagons? One possible solution is to expand the alphabet to 25 letters and, as Borges did, include the space, the comma, and the period to round the total up to $28 = (2^2) \cdot 7$ orthographic symbols. Then, if the other (admittedly arbitrarily chosen) numbers for each book stay the same, there will be $28^{1,312,000}$ distinct books.

Next, hire infinitely many cabinetmakers to rebuild the bookshelves in the hexagons, so that each of the four walls holds four shelves, and each shelf holds 49 books. Then a total of $4 \cdot 4 \cdot 49 = 784 = (2^4) \cdot (7^2)$ books furnish each hexagon, and since

$$\begin{aligned} \frac{28^{1,312,000}}{784} &= \frac{(2^{2,624,000}) \cdot (7^{1,312,000})}{(2^4) \cdot (7^2)} \\ &= \frac{(2^4) \cdot (2^{2,623,996}) \cdot (7^2) \cdot (7^{1,311,998})}{(2^4) \cdot (7^2)} \\ &= (2^{2,623,996}) \cdot (7^{1,311,998}), \end{aligned}$$

after the renovation, the $28^{1,312,000}$ books exactly fill $(2^{2,623,996}) \cdot (7^{1,311,998})$ hexagons.

For this last section, the aim is to explain concisely why we are currently, and for the foreseeable future, unequal to the task of determining the median of the prime numbers expressible in 100 digits. The median of the set of primes expressible in 100 digits is, in a sense, the "middle" of all of those primes. To compute the median, arrange the numbers sequentially from the smallest to the largest prime less than 10^{100} (which is called one googol).

$$2, 3, 5, 7, \ldots \text{(about } 10^{97} \text{ more primes)} \ldots, \underbrace{9999\ldots999}_{97 \text{ digits of nine}}203.$$

Now, if there are an odd number of primes in the list, the median is the absolute middle of the list. If there are an even number of primes in the list, the median is the average of the two primes appearing in the middle of the list. (The average of these two numbers is guaranteed to be an integer, for the sum of two odd numbers is even, and we conclude the calculation of the average by dividing by two.)

The only way to find the median would be, in one way or another, to account for the complete list of prime numbers expressible in 100 digits. Including 0, there are exactly one googol numbers expressible in 100 digits. By the famous prime number theorem—which we'll outline in a moment—there are more than 10^{97} prime numbers smaller than 10^{100}. This number may sound manageable, but 10^{97} is trillions of times larger than the number of subatomic particles in our universe. There simply isn't any imaginable way to list and keep track of 10^{97} numbers, which precludes the possibility of finding the median.[1]

The prime number theorem was first conjectured in various forms by Euler and others beginning in the late eighteenth century and was finally proved about a hundred years later in 1896 by Hadamard (and independently that same year by Poussin). Part of the beauty of the prime number theorem is that it provides an excellent estimate of how many primes there are that are smaller than 10^{100} without explicitly naming a single one!

The prime number theorem says that if $\pi(n)$ is equal to "the number of primes less than or equal to n," then as n grows very large,

$$\pi(n) \approx \frac{n}{\ln(n)},$$

where $\ln(n)$ is the natural log function. (The natural log has the same remarkable properties as the log function, $\log(n)$, that we looked at earlier, and indeed, after multiplication by a constant, they are the same function.) We are interested in knowing approximately the number of primes expressible in 100 digits, so we compute $\pi(10^{100})$ for a good estimate:

$$\pi\left(10^{100}\right) \approx \frac{10^{100}}{\ln\left(10^{100}\right)} = \frac{10^{100}}{100 \cdot \ln(10)} = \frac{10^{98}}{\ln(10)} \approx \frac{10^{98}}{2.3} \approx 10^{97}.$$

THREE

Real Analysis
The Book of Sand

> *To see a World in a Grain of Sand*
> *And a Heaven in a Wild Flower*
> *Hold Infinity in the palm of your hand*
> *And Eternity in an hour.*
> —William Blake, "Auguries of Innocence"

REAL ANALYSIS IS THE BRANCH OF MATHEMATICS that explores, among other ideas, the nuances of the arbitrarily small. Paradoxically, in this chapter, thinking about the very small will prove decisive in understanding the very large: the Book that embodies the entire Library.

Borges' last insight regarding the Library is cloaked in a footnote adorning the conclusion of the last sentence. The footnote reads:

> Letizia Álvarez de Toledo has observed that the vast Library is pointless; strictly speaking, all that is required is *a single volume*, of the common size, printed in nine- or ten-point type, that would consist of an infinite number of infinitely thin pages. (In the early seventeenth century, Cavalieri said that every solid body is the superposition of an infinite number of planes.) Using that silken *vademecum* would not be easy: each apparent page would open into other similar pages; the inconceivable middle page would have no "back."

Others have independently noticed that Borges continued to play with the idea of such a Book in his evocative short story "The Book of Sand."[1]

The mathematical analysis of a Book of Sand hinges on what is meant by the phrase "infinitely thin pages." Three different interpretations of "infinitely thin" lead to three Books similar in spirit, but disparate in the details. We'll examine them in ascending order of exoticness.

First Interpretation

If we take "infinitely thin" to mean merely "thinner than any subatomic particle," there are several refreshing possibilities. First, there are $(410) \cdot \left(25^{1,312,000}\right)$ pages in the Library, a very large number, but still finite. Thus, if every page is the same thickness, say

$$\frac{1}{(410) \cdot \left(25^{1,312,000}\right)} \text{th of an inch,}$$

then the Book, sans cover, will be exactly one inch thick. Such a Book, though, would defraud the anonymous librarian of his "elegant hope" that the Library is repeated in its disorder, and also contravene the explicit statement in the footnote that the book would consist of an infinite number of pages. If, as above, the pages were all the same thickness, then an infinite periodic repetition of all the books of the Library would force the Book of Sand to be infinitely thick.

If we insist on each page having a definite thickness, and we equally insist upon infinite repetitions for the pages of the Book, we must therefore allow for ever-thinner pages. To make sense of such a Book, we need to understand an idea from the theory of infinite sums.[2] We'll begin this short journey by treading parallel to the tiny footfalls, echoing loudly through the ages, of the Paradox of Zeno so beloved by Borges.

Suppose, starting at one end of a room, we were to walk halfway across towards the opposite wall. After a brief pause, we walk half the distance from the midpoint towards the opposite wall. After another brief pause, we walk half the distance... (see figure 5).

In the coarse world we inhabit, we'll stub our toes on the wall in short order. In the idealized world of mathematics, we may always halve the distance between one point and an endpoint. (Zeno's and Parmenides' paradoxes exploit this chasm between the world of our perceptions and the mathematical vision of a line segment.)

For the purposes of this book, without offering a rigorous proof, note that by adding up the lengths symbolized by the arcs, the information

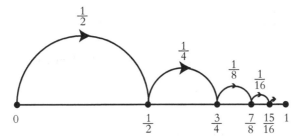

FIGURE 5. Zeno walks across the room from 0 to 1.

encoded in figure 5 is equivalent to this equation:

$$\frac{1}{2} + \frac{1}{4} + \frac{1}{8} + \frac{1}{16} + \frac{1}{32} + \cdots = 1.$$

This equation encapsulates a striking fact: by adding up infinitely many segments, each smaller than the previous by one-half, a form of unity is achieved. A gauge of the depth and profundity of this insight is that for several centuries, most thinkers conceded that this hammered home the final nail in the coffin for Zeno's Paradox. (Nowadays, thinkers have again complexified the picture, thereby casting doubt, raising questions, and essentially resurrecting the dead.)

Following the example set by the equation, choose the first page to be one-half of a standard page's thickness, then the next page half that thickness, the next half that thickness, and so on and so on. Then the entire Book, infinitely periodically repetitive, will be exactly one standard page thickness.

In the Math Aftermath following the chapter, we provide a bit more background on this next calculation, which is estimating the thickness of the 41st page. We conclude that the 41st page is

$$\underbrace{\left(\frac{1}{2}\right) \cdot \left(\frac{1}{2}\right) \cdot \left(\frac{1}{2}\right) \cdot \cdots \cdot \left(\frac{1}{2}\right)}_{\text{Cut the first page in half 40 times.}} \times \text{(one standard page thickness)}$$

$$= \left(\frac{1}{2}\right)^{40} \times \text{(one standard page thickness)}$$

$$\approx \left(\frac{1}{1{,}099{,}511{,}627{,}776}\right) \cdot \left(\frac{1}{1{,}000} \text{ meter}\right)$$

$$\approx \left(\frac{1}{10^{12}} \cdot \frac{1}{10^3}\right) \text{ meters} = \left(10^{-12}\right) \cdot \left(10^{-3}\right) \text{ meters}$$

$$= 10^{-15} \text{ meters thick,}$$

which is thinner than the diameter of a proton. Since each successive page is one-half the thickness of the preceding page, all the rest of the pages are also thinner than a proton. Of course, in this interpretation, though almost every page is invisible to the naked eye, or even an electron microscope, *it is not the case that any page is actually "infinitely thin."*

Second Interpretation

Here, we take "infinitely thin" in the sense indicated by the reference to Cavalieri's principle in the footnote: the thickness of a Euclidean plane. The thickness of a plane is the same as the length of a point, which is tricky to define. Consider a point in the line. It is clear that a Euclidean point is thinner than a line segment of *any* positive length. It is somewhat disquieting, though, to say that a point has length 0; if so, how does massing together sufficiently many 0-length entities create a line of positive length? Doesn't adding together 0s always produce another 0? How could an object be of length 0?

A subtle way of evading these traps was crafted at the beginning of the twentieth century, primarily through the work of Henri Lebesgue, whose theory is now a vast edifice with ramifications permeating much of modern mathematics. Fortunately, we need only a small cornerstone of the theory: the idea of a set of measure 0 contained in the *real number line*.

Recall that the real number line consists of all rational and irrational numbers, each representing a point on the line, each also signifying the distance from the origin to the point. It may be confusing that we are explicitly identifying the "length of an interval" with a "number," for again, a real-world idea, that of length, is interpenetrating a mathematical idealization. We inhabit this limbo for the rest of the chapter.

We need two definitions. A *closed interval* includes both endpoints of an interval; as an example, the notation [0, 1] means "all numbers between

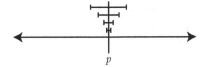

FIGURE 6. An arbitrarily small interval may contain the point p.

0 and 1, inclusive." Now, let **S** be any set contained in the real number line. One says that **S** is *a set of measure 0* if **S** can be contained in a union, possibly infinite, of closed intervals whose lengths add up to an arbitrarily small number. Several examples will help clarify this definition.

Example 1. A single point p in the real number line. Clearly p can be contained in a closed interval of arbitrarily small length (figure 6). Thus p is a set of measure 0. Note the fine distinction: we are not saying "the point p is of length 0"; rather we are saying that p is a set whose measure is 0. It turns out—and we'll see an example soon—that there are sets of measure 0 which are quite counterintuitive.

Example 2. Three points a, b, and c in the real number line. Let

a be contained in an interval of length 1/2,
b be contained in an interval of length 1/4, and
c be contained in an interval of length 1/8.

(It doesn't matter if the intervals overlap.) Then the three points are contained in a union of intervals whose sum-length is

$$\frac{1}{2} + \frac{1}{4} + \frac{1}{8} = \frac{7}{8}.$$

Not arbitrarily small yet! But now, let

a be contained in an interval of length 1/4,
b be contained in an interval of length 1/8, and
c be contained in an interval of length 1/16.

Then, since each interval is half the length of its corresponding predecessor, the sum is also halved.

$$\frac{1}{4} + \frac{1}{8} + \frac{1}{16} = \frac{7}{16}.$$

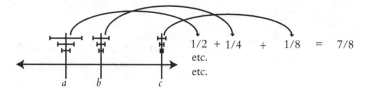

FIGURE 7. Here, we sum ever-smaller triples of intervals. The sum of each triple is half the length of the preceding triple's sum.

If we play this game again, starting with an interval of length 1/8, we find that

$$\frac{1}{8} + \frac{1}{16} + \frac{1}{32} = \frac{7}{32}.$$

If we continue to put a, b, and c in intervals half of the lengths of the previous go-round, the triple of intervals will also sum to half the preceding length: first 7/64, then 7/128, and so on. By starting with a sufficiently small interval, we ensure the sum of the three intervals is arbitrarily small—that is, the set $\mathbf{S} = \{a,b,c\}$ is a set of measure 0 (figure 7).

Example 3. It is a curious fact that it is difficult to show that the interval of numbers between 1 and 4 is *not* of measure 0. Certainly our intuition informs us that the minimum length of intervals necessary to cover [1, 4] will sum to 3, but demonstrating it rigorously is a nontrivial exercise, well beyond the scope of this book. See figure 8.

Back to the infinitely thin pages of the Book. We interpret "infinitely thin" as meaning that each page has a thickness of measure 0. We also assume, as we did at the end of the first interpretation, that within this Tome, the books of the Library repeat over and over, enacting the

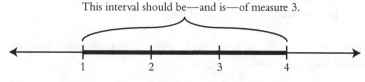

FIGURE 8. The measure of the interval from 1 to 4.

anonymous librarian's "elegant hope" of a periodically repeating order. We are therefore confronted with an intriguing problem:

There are infinitely many pages, each of which has thickness of measure 0. How thick is the Book?

The answer might run counter to your intuition:

The thickness of the Book is of measure 0.

In other words, if we looked at the Book sideways, we would not be able to see it, let alone open it. How does this unexpected, unimagined, unimaginable state of affairs arise? Once we think to look for it, it turns out to be sitting there, almost as if it was waiting to be discovered.

The goal is to show that the thickness of the Book can be contained in a collection of closed intervals which can be chosen so that the sum of their lengths can be made arbitrarily small. If this can be done, then by definition the Book is of measure 0. We'll accomplish this by covering the thickness of each page in ever-smaller intervals in a sneaky way that exploits the infinite sum that embodied Zeno's Paradox.

First, though, another counterintuitive point, followed by a technical one. Although it's conceivable that the Bookbinder bound the infinitely many pages of the Book together in a straightforward order, it is also possible that the pages of the Book wash up against themselves similar to the *rational numbers*, meaning there is no more a "first" page of the Book than there is a "first" positive rational number. If so, we simply choose one of the $25^{1,312,000}$ books to be the first, another to be the second, and so on, until we have a complete list of the books and their pages. Since the Book repeats, we are thus able to give numbers to its pages.[3] So, let

the first page be contained in an interval of length 1/2,
the second page be contained in an interval of length 1/4,
the third page be contained in an interval of length 1/8,
and so on,
and so on.

We saw in the first interpretation that

$$\frac{1}{2} + \frac{1}{4} + \frac{1}{8} + \frac{1}{16} + \frac{1}{32} + \cdots = 1,$$

so the thickness of the Book can contained in an infinite union of intervals which sum to 1. Here's where the sneaky part comes in. Now, let

> the first page be contained in an interval of length 1/4,
> the second page be contained in an interval of length 1/8,
> the third page be contained in an interval of length 1/16,
> and so on,
> and so on.

This time, the infinite union of intervals sums to

$$\frac{1}{4} + \frac{1}{8} + \frac{1}{16} + \frac{1}{32} + \frac{1}{64} + \cdots = \frac{1}{2}.$$

This is seen by simply subtracting 1/2 from both sides of the previous equation. Notice how we are exploiting an aspect of the idea of infinity: we are throwing away a term from the left side of the equation, but still have infinitely many terms to account for the infinite number of pages.

If we start by letting the thickness of the first page be contained in an interval of length 1/8, then the sum becomes:

$$\frac{1}{8} + \frac{1}{16} + \frac{1}{32} + \frac{1}{64} + \frac{1}{128} + \cdots = \frac{1}{4}.$$

Clearly, by continuing to play this game of lopping the intervals in half, we ensure that we may always find a union of intervals that contains the thickness of the Book and sums to an arbitrarily small number. This means that the thickness of the Book is of measure 0, an outcome surely unimagined by Borges.

How it is possible to create a line segment, a set of positive measure, out of points of measure 0? That is a long story for another day.

Third Interpretation

Perhaps the elusive nature of the preceding interpretation is unsatisfying; we never took a stand on how thick "infinitely thin" is; we merely observed that it is of measure 0. For the third interpretation, we will glimpse some of the basic elements of one of the most underutilized mathematical theories of the twentieth century: nonstandard analysis. The roots of the development of nonstandard analysis began with Leibniz, one of the inventors of the calculus. Both Leibniz and Newton used infinitely small quantities, *infinitesimals* (also known as *fluxions*), in their early calculations. In his foreword to the revised edition of Abraham Robinson's seminal work, *Non-standard Analysis*, the logician Wilhelmus Luxemburg notes that "Bishop Berkeley disdainfully referred to infinitesimals as the 'ghosts of departed quantities,'" and that in response to this and other attacks, "Leibniz proposed a program to conceive of a system of numbers that would include infinitesimally small as well as infinitely large numbers."

Because of the difficulties inherent in beginning Leibniz's bold program, and for other historical reasons, his ideas lay fallow for almost 300 years. In 1961, with the publication of *Non-standard Analysis*, Robinson rebutted Berkeley and fulfilled Leibniz's dream. Using various tools of logic and set theory developed in the late nineteenth and early twentieth centuries, Robinson was able to create a consistent, logical model of a number system that included infinitesimals.

It should be mentioned, with sincere respect, that adherents of nonstandard analysis possess a striking combination of mystic fervor and matter-of-fact pragmatism about the topic. This may be because the mainstream of mathematics has, at least for now, marginalized nonstandard analysis due to its less intuitive constructions and technical complexities. Bearing this in mind, here are selections, originally excerpted by Mark McKinzie and Curtis Tuckey, from H. Jerome Keisler's college textbook, which approaches the calculus from the nonstandard viewpoint (emphases added by present author).

> In grade school and high school mathematics, the real number system is constructed gradually in several stages. Beginning with the positive integers, the systems of integers, rational numbers and finally real numbers are built up...

> What is needed [for an understanding of the calculus] is a sharp distinction between numbers which are small enough to be neglected and numbers which aren't. Actually, no real number except zero is small enough to be neglected. *To get around this difficulty, we take the bold step of introducing a new kind of number, which is infinitely small and yet not equal to zero* . . .
> The real line is a subset of the hyperreal line; that is, each real number belongs to the set of hyperreal numbers. Surrounding each real number r, we introduce a collection of hyperreal numbers infinitely close to r. *The hyperreal numbers infinitely close to zero are called infinitesimals.* The reciprocals of nonzero infinitesimals are infinite hyperreal numbers. The collection of all hyperreal numbers satisfies the same algebraic laws as the real numbers . . .
> We have no way of knowing what a line in physical space is really like. It might be like the hyperreal line, the real line, or neither. However, in applications of the calculus it is helpful to imagine a line in physical space as a hyperreal line. The hyperreal line is, like the real line, a useful mathematical model for a line in physical space.

In nonstandard analysis, there are infinitely many hyperreal infinitesimals clustered around 0, every one smaller than any positive real number. Each signifies an infinitely small distance. We may simply assign any infinitesimal we wish to each page of the Book.[4] By the rules of nonstandard analysis, we compute the thickness of the Book by adding together all of the infinitesimals. For a summation such as this one, adding the infinite number of infinitesimals produces yet another infinitesimal, so the Book is, again, infinitely thin: never to be seen, never to be found, never to be opened. This time, though, we may elegantly console ourselves that the infinite thinness is a precisely calculable nonstandard thickness.

> Regardless of which interpretation we assume, if the pages are 'infinitely thin,' then by necessity the Book of Sand itself is infinitely thin.

Math Aftermath: Logarithms Redux

> *Reason looks at necessity as the basis of the world; reason is able to turn chance in your favor and use it. Only by having reason remain strong and unshakable can we be called a god of the earth.*
> —Johann Wolfgang Von Goethe, *Wilhelm Meister's Apprenticeship*, bk. I, ch. 17

Recall that in the first Math Aftermath, we used logarithms to solve an equation involving exponentials. This is another example, only slightly more complicated, of using logarithms to solve an equation. Earlier in this chapter, we claimed that if the Book of Sand started with a normal page thickness, say one millimeter, 10^{-3} meters, and each successive page was half the thickness of the preceding page, then the 41st page would be thinner than a proton, which measures a little more than 10^{-15} meters across. How did we find the number 40?

Let's set it up as an equation. Each page is half the thickness of the preceding page, so if we measure the nth page after the first page, it will be the thickness of the first page cut in half n times. That is, it will be

$$\frac{10^{-3}}{2^n} \text{ meters across.}$$

Since the size of the proton is approximately 10^{-15}, we set these two terms equal to each other and then simplify the equation.

$$\frac{10^{-3}}{2^n} = 10^{-15}, \text{ which implies } \frac{10^{-3}}{10^{-15}} = 2^n;$$

$$\text{therefore, } 10^{12} = 2^n.$$

Solving this last equation without logarithms would be very difficult. (In fact, in 2004, powerful mathematical software running on my late-model computer crashed the computer in a failed, naïve, brute-force attempt to solve for such an n.) Since 10^{12} and 2^n, although written differently, are the same number, it should again be the case that any function applied to both of them will output the same number. Thus,

$$log\left(10^{12}\right) = log\left(2^n\right),$$

which, by using the remarkable property of the logarithm, entails that

$$12 \cdot log\left(10\right) = n \cdot log\left(2\right).$$

Dividing both sides by $log(2)$ yields

$$\frac{12 \cdot log\left(10\right)}{log\left(2\right)} = n,$$

which can quickly be solved with a computer, a calculator, or—for traditionalists—logarithm tables. When we do so, we find that n is about equal to 39.9, so to ensure we get the result we want, we round upwards. Thus, if we cut the initial page's thickness in half 40 times, it will be the case that the 41st page is thinner than a proton.

FOUR

Topology and Cosmology
The Universe (which Others Call the Library)

> *A fact is the end or last issue of spirit. The visible creation is the terminus or the circumference of the invisible world.*
> —Ralph Waldo Emerson, "Nature"

TOPOLOGY IS A BRANCH OF MATHEMATICS THAT explores properties and invariants of spaces, and for the purposes of this book we consider a space to be a set of points unified by a description. Cosmology is quite literally the study of our cosmos. If we consider the Library to constitute a universe and the universe to be the Library, it is not unreasonable to combine these notions and speculate as to a conceivable topology of the Library that best reflects the anonymous librarian's received wisdom and secret hopes.

Early in the story—and many commentators have noted the connection between the italicized phrase and Borges' essay "Pascal's Sphere"—Borges writes

> Let it suffice for the moment that I repeat the classic dictum: *The Library is a sphere whose exact center is any hexagon and whose circumference is unattainable.*

The final sentences of the story invite us to reopen the question of the topology of the Library:

> I am perhaps misled by old age and fear, but I suspect that the human species—the only human species—teeters at the verge of extinction, yet that the Library—enlightened, solitary, infinite,

perfectly unmoving, armed with precious volumes, pointless, incorruptible, and secret—will endure.

I have just written the word "infinite." I have not included that adjective of out of mere rhetorical habit; I hereby state that it is not illogical to think that the world is infinite. Those who believe it to have limits hypothesize that in some remote place or places the corridors and stairs and hexagons may, inconceivably, end—which is absurd. And yet those who picture the world as unlimited forget that the number of possible books is *not*. I will be bold enough to suggest this solution to the ancient problem: *The Library is unlimited but periodic.* If an eternal voyager should journey it in any direction, he would find after untold centuries that the same volumes are repeated in the same disorder—which, repeated, becomes order: the Order. My solitude is cheered by that elegant hope.

Collecting the properties of the classic dictum (CD) and the Librarian's solution (LS), we obtain the following list:

1. Spherical (CD)
2. Center can be anywhere—uniform symmetry (CD)
3. Circumference is *unattainable*. (CD)
4. No boundaries (LS)
5. *Limitless* (LS)
6. Periodic (LS)

Is there a space that embodies all six of these properties? If so, how can we best envision it and grasp it with our intellect? We claim there is an excellent candidate that encompasses these properties, if we are willing to refine our interpretations just a smidge. In the Math Aftermath to this chapter, we discuss two other compelling ways of configuring the Library that each significantly expand our conceptions of the possible.

Let's begin with the space most familiar to our intuitive geometric sense: Euclidean three-dimensional space (henceforth, *3-space*). It is a space we think of as possessing volume, as having three axes of orientation with ourselves as the central point; we may move forward or backwards, we may move left or right, and we may move up or down. And, of course, we may also move in combinations of these directions. Notice that from this description, there is no fixed preferred center point: *we are our own central points.*

Indeed, one of Descartes' deepest ideas was to specify a point—some point, any point—in 3-space and call it the origin. Three axes intersecting at the origin, typically called the x, y, and z axes, are set with each axis at right angles to the other two. They abstract our innate, intuitive orientation and, with the introduction of a unit length, which naturally induces a numbering of the axes, give rise to a coordinatization of space. Algebra can now conjoin geometry, creating analytic geometry, and later spawn calculus.

But there are no distinguished points of any kind in Euclidean 3-space; in fact, the view from any point is the same as from any other point. There are no walls, no boundaries, and no limits. It seems at the end of the story the librarian envisioned this kind of space, partitioned into hexagons, filled with books, extending infinitely throughout the totality of 3-space. The books' shelving pattern repeats endlessly along each of the three axes, much as a symmetric wallpaper pattern does in two dimensions. While this conception of the Library satisfies points 2, 4, 5, and 6, it also induces a vertiginous disorientation born of trying to imagine a thing extending away forever. For example, if the Library goes *down* forever, what do the hexagons rest on? More hexagons? Rather remarkably, the architectural, model of the Library that we propose provides a satisfying answer to this question.

A note regarding the gravity of the situation. If the universe and the Library are synonymous, and if we make the reasonable assumption that the universe is neither expanding nor contracting, it follows that the natural gravitational field would be identically zero everywhere. Even though there are unimaginable amounts of matter in the universe/Library, its homogeneous distribution entails that the gravitational effect from any one direction would be canceled out by precisely the same effect from the opposite direction. Since the builders of a Library must be, at least from our perspective, omnipotent, their talents surely must include the ability of imposing a useful constant gravitational field on the Library.

Euclidean 3-space embodies some of the qualities of interest in our quest to understand the large-scale structure of the Library. We need to limn two more ideas, one mathematical, one mystical, before we can describe the form of a Library that reconciles the characteristics of the classic dictum and the Librarian's solution.

The mathematical idea is relatively recent—it comes from the early part of the twentieth century. For the purposes of this book, we'll say that a *manifold* is a space that is *locally Euclidean* but that on a global scale may be *non-Euclidean*. Perhaps the simplest possible example is that of a sphere, or globe, or surface of a cantaloupe, or of the earth, balloon, soccer ball; take your pick. Locally, assuming that we are so small we can't detect the curvature, each micropatch of a sphere is, in essence, a two-dimensional Euclidean plane (*2-space*). One need think only of the steppes of Central Asia, the corn belt of the United States, the Sahara desert, or any large, calm body of water to engage vivid testimony on this point. Globally, despite the essential flatness of each little patch, we find non-Euclidean behavior: if we begin at a point, pick a direction, and continue moving in that direction, we circumscribe the sphere and return to our starting point. This can't occur in 2-space, where we perforce travel forever in one direction and can't ever come close to a previously visited point.

Again, a manifold is *locally* Euclidean. If we start at any point in space, look around and take a few steps in any direction, do we think we are in Euclidean space? If the answer is yes, then we are in a manifold. If we continue walking, and some unusual phenomenon occurs, such as returning to our starting point, then we realize we are in a nontrivial manifold; that is, one with global non-Euclidean properties. Our universe, for example, seems to be a manifold, although interesting questions arise at black holes. Certainly one cannot imagine standing at a black hole and taking a step in any direction! Researchers are trying to devise methods of determining the global structure; a readable introduction to this area of research can be found in Luminet et al.

The mystical idea is relatively ancient—I leave it to a Borgesian intellect to trace its roots and agelong echoes. Let's begin in a familiar place, our own universe. If we talk about an object in our universe— for example, a desk or chair—we view it as embedded in a larger space. Consequently, we often use our relative coordinate system to refer to objects, as when we say "It's on my right," or "Over there! Directly behind you, to the left," or "Scratch my back...lower...lower...to the right...now up...that's it!" Over the millennia, primarily as navigation aids, we've settled upon somewhat less arbitrary reference points, such as the North Star, the magnetic North Pole, and the true North Pole. The point is, though, that these references, these origins, are all within our universe. "Outside the universe" is a phrase beyond normal comprehension. Some theories place our universe in a larger matrix, such as a superheated gas cloud containing an infinite number of inaccessible universes, or in a higher-dimensional space, or in a multiplicitous welter of bifurcating universes. However, these theories raise the question,

"What is outside of the larger universe?"

Really, now, though, "What is outside of the universe?" The answer is no thing; nothing; non-space; indescribability; un-thing-ed-ness; Void beyond vacuum: all these non-things are the "outside" of our universe.

These two ideas, the mathematical and the mystical, are woven together in this question and its answer.

Where is the center of a sphere?

If the sphere is considered as an everyday object embedded in our universe, the answer may take a form such as "at the intersection of two diameters," or, pointing at it dramatically, saying with particular emphasis, "There! In the middle, in the interior!" See figure 9. If, though, we consider the surface of the sphere as a manifold, as a space in itself and of itself, then the question and answer are subtler. As in the case of our universe, as if we were points residing in the sphere itself, there is no legitimate referral to a point outside the universe of the surface of the sphere. There is only the sphere; every thing else is no thing. Where is the center of a sphere? Considered as a manifold, then, the answer is

Everywhere and nowhere.

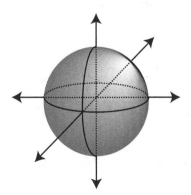

FIGURE 9. Where is the center of a sphere?

Every point has the property that locally, it looks like Euclidean space, and regardless of the direction taken, consistently moving in any chosen direction returns us to the starting point. No point is distinguished in any way.

∽

One more idea is necessary to provide a satisfying topology for the Library. The example of a manifold we used was a two-dimensional sphere (*2-sphere*). There are a number of ways to rigorously define a 2-sphere. Euclid might write something like, "Given a point p in 3-space, a sphere with center p is the collection of all points a specified uniform distance away from p." An analytic geometric equation for the standard unit sphere is $x^2 + y^2 + z^2 = 1$. (If you're interested in seeing why this equation specifies a sphere, please turn to the appendix "Dissecting the 3-Sphere".) Here, using words and pictures, we provide a topological construction of a 2-sphere.

Start with a disk in the Euclidean plane and while preserving the interior of the disk except for bending and stretching, crimp the entire boundary circle up out of 2-space, and then contract the boundary to one point. This point, the contraction of the boundary, becomes the north pole and vanishes into the surface of the sphere created as the process is completed (figure 10). An interesting point: the way we've described it, and the way the picture shows this process, it seems as though a disk is being modified over time. By contrast, though, one should simply say, "Identify the boundary of the disk to a point." Thus, in some sense, the creation of the sphere is a timeless step that happens "all at once."

FIGURE 10. A disk curls up out of the Euclidean plane, while its interior stretches and its boundary circle shrinks to a point. The result is a 2-sphere.

The three-dimensional sphere (*3-sphere*) provoked many advances in topology over the past century, and due to the recently solved Poincaré conjecture it remains a vibrant research topic. The 3-sphere is a generalization of the *1-sphere*—a circle—and the 2-sphere. Euclid might write something such as, "Given a point p in *four*-dimensional Euclidean space, a sphere with center p is the collection of all points a specified uniform distance away from p." An analytic geometric equation for the standard unit 3-sphere is $w^2 + x^2 + y^2 + z^2 = 1$. (Again, to see how this equation captures "sphericality," please consult the appendix "Dissecting the 3-Sphere.") An analogous topological construction for the 3-sphere is difficult to envision, but by pushing the limits of our understanding, we may learn much.

Take a solid ball—a baseball, or an apple, or a cherry, or a cannonball—and, while leaving the interior of the ball uncompressed, crimp the entire boundary sphere upwards, and then simply contract the boundary sphere to one point (figure 11). That's it. At least the *difficulty* is easy to understand; for the construction of the 2-sphere, we took a two-dimensional object, the disk, and had to bend it into the third dimension before we could contract the boundary at all. Starting with a solid ball in three dimensions, we must "bend" the ball into the fourth dimension before we can contract the boundary (figure 12). At this juncture, the mathematics becomes unimaginable; the best to be hoped for is that by meditating on the lower-dimensional examples accessible to our imagination, we may be able to conjure the memory of the trace of a once-sensed intuition. Still, by proceeding with analogies to the 2-sphere, we'll use a trio of methods to begin to visualize the 3-sphere.

If we take a two-dimensional Euclidean slice of a 2-sphere, the resulting geometric object is either a point—at the north and south poles—or a

FIGURE 11. Trying to shrink the spherical boundary of a solid ball to a point without shrinking the whole ball. It can't be done in 3-space.

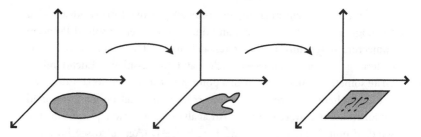

FIGURE 12. The analogous problem for a disk. Confined to 2-space, the boundary of the disk can't be contracted to a point without shrinking the entire disk.

1-sphere (figure 13). Using a mild updating of an idea from *Flatland*, if we make a movie of the slice moving from the north pole to the south pole, a viewer would see a point that grows into a unit circle, which then shrinks back down to a point (figure 14, left). In a similar fashion, if we take a three-dimensional Euclidean slice of a 3-sphere, the resulting geometric object is either a point—at the "top" or "bottom"—or a 2-sphere. If we make a movie of the slice moving from the top to the bottom, the viewer would see a point that grows into a unit sphere, which then shrinks back down to a point (figure 14, right). (Again, for those who find equations more convincing than pictures, we provide an analytical proof of this in the appendix "Dissecting the 3-Sphere.")

Expanding on this idea, suppose we were forced to squish the 2-sphere, whose natural home is in 3-space, down into 2-space. Since we just conceived of the 2-sphere as a collection of stacked circles combined with two poles, we may envision a flattened planar depiction as a

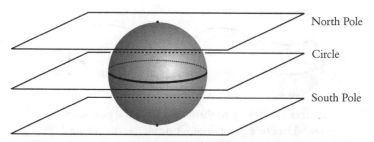

FIGURE 13. Taking slices of a 2-sphere by Euclidean planes.

TOPOLOGY AND COSMOLOGY 65

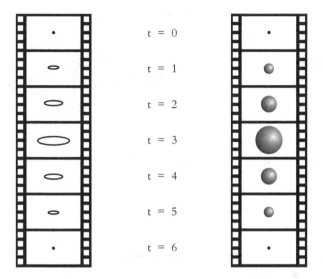

Time-lapse film of a
Euclidean plane passing
through a 2-sphere.

Time-lapse film of a
Euclidean 3-space passing
through a 3-sphere.

FIGURE 14. A film of planar slices of the 2-sphere and volume slices of the 3-sphere.

collection of intersecting circles with two points signifying the north and south poles (figure 15). The related problem, the one that's been tasking us, is how to represent the 3-sphere down-sized into 3-space. If we think of the 3-sphere as "stacked" 2-spheres—in the same sense that a 2-sphere is stacked 1-spheres—the analogous 3-space representation is a collection of intersecting 2-spheres (figure 16).

For the third way of envisioning the 3-sphere, the lower-dimensional correlate is to take a section of the 2-sphere and flatten it out into the Euclidean plane. If our section includes, say, the south pole, the flattened section is a disk. If our section doesn't include either pole, the flattened section is an *annulus*, which is a ring, a thickened circle. Note that the equator of the sphere (the dotted circle in figure 17) is flattened to the central circle of the annulus. The circle-slices above the equator on the sphere are smaller than the equator, but when flattened become *larger* than the central circle of the annulus. Similarly, the circle-slices below

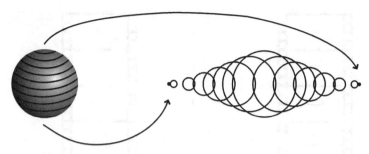

FIGURE 15. The circles and two points of a 2-sphere flattened into the plane.

FIGURE 16. The "flattening" of a 3-sphere into 3-space is analogously represented by intersecting 2-spheres combined with two points.

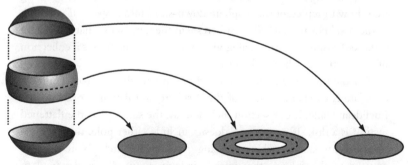

FIGURE 17. Flattened sections of the 2-sphere are either disks or annular rings.

TOPOLOGY AND COSMOLOGY   67

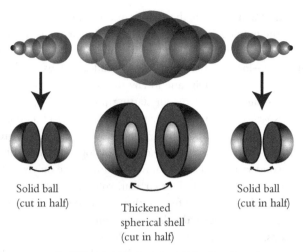

FIGURE 18. "Flattened" sections of the 3-sphere in 3-space are either solid balls or thick spherical shells.

the equator flatten to even smaller circles in the annulus than they were in the sphere. This process of dimensional flattening distorts the object; necessarily, information is lost.

If we take sections of the 3-sphere, we must consider how to "flatten out" the resulting object into 3-space. If our section includes, for example, the bottom of the 3-sphere, the flattened section is, by analogy, a solid ball. If the section of the 3-sphere doesn't include the north or south poles, the "flattened" section is a solid ball with a smaller ball removed from the center—a pitless olive, or a tennis ball, or an empty walnut shell, or a thickened spherical shell. In figure 18, perhaps the most counterintuitive aspect is the means by which the middle collection of 2-spheres collates to a thickened spherical shell. The centralmost, the largest 2-sphere, is flattened to itself. The smaller spheres directly to the left, say, of the central sphere thicken it on the inside. The smaller spheres directly to the right of the central sphere are distorted by the flattening into *larger* spheres that thicken the exterior of the central sphere. Again, unfortunately, the process entails that we must lose information about the size of the spheres.

All the girders and struts of the framework are now in place to finish assembling the topology and cosmology of the Library. The 3-sphere is a three-dimensional manifold; at every point, if we inhabited the 3-sphere, we would say—locally—that space was Euclidean. If we walked what we perceived to be a straight line in any direction, we would—possibly after countless ages—return to our starting point; the 3-sphere can be construed as *periodic*. There are *no boundaries*, no walls to bump into; the 3-sphere is *limitless*. Moreover, in his luminous story "The Garden of Forking Paths," Borges has the sympathetic sinologist Stephen Albert say, "... I had wondered how a book could be infinite. The only way I could surmise was that it be a cyclical, or circular, volume, a volume whose last page would be identical to the first, so that one might go on indefinitely." Even though Albert rejects this line of reasoning for "The Garden of Forking Paths," this quote, coupled with Borges' well-known interest in Nietzsche's idea of eternal recurrence, indicates that Borges was willing to consider cyclic or recurrent structures as tokens of, or synonymous with, infinity.

Considered as a three-dimensional manifold, *the center of the 3-sphere is everywhere and nowhere*. Furthermore, if the 3-sphere is so large that, regardless of our transport, we could never come close to circumnavigating it, it would not be illegitimate to say that the *circumference is unattainable*. Finally, this answers the question concerning what the hexagons "rest on." By forming *great circles*—circles which are essentially equators of a sphere—the hexagons all rest upon each other and ultimately themselves, and thus there is no need for an impossible "external" foundation for the Library.[1]

Now, though, it's conceivable that generalizing from a 2-sphere might generate some disquietude: on a 2-sphere, any two distinct great circles intersect at exactly two points (figure 19). It is not unreasonable to worry

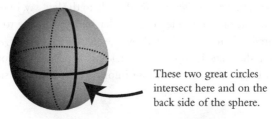

These two great circles intersect here and on the back side of the sphere.

FIGURE 19. A possible source of unease.

that any two distinct great circles on the 3-sphere would also of necessity intersect in at least two points. This might entail that all the air shafts and all the spiral staircases would converge, say, at the north and south poles of the 3-sphere, causing a traffic jam of epic proportions. Fortunately, this intuitively plausible scenario doesn't happen. Perhaps the easiest way to begin to get a handle on why this isn't a problem is to grasp that a circle is only *one* dimension smaller than a 2-sphere. Consequently, it has special properties of "dividing" space locally into two pieces; certainly a great circle divides a 2-sphere into two hemispheres. However, a circle is *two* dimensions smaller than a 3-sphere and hence has no such special division property in the 3-sphere. Imagine a circle floating in the center of the room—space flows through it and around it with aplomb.

If the Library is the universe, and the universe is a 3-sphere, then *the Library is a sphere whose exact center is any hexagon and whose circumference is unattainable; moreover, it is limitless and periodic.* That is, the 3-spherical Library satisfies both the classic dictum and the librarian's cherished hope.

Math Aftermath: Flat Out Disoriented

The reverse side also has a reverse side.
—Japanese proverb

Donuts. Is there anything they can't do?
—Homer Simpson, *The Simpsons*

The enemy of my enemy is my friend.
—Ancient proverb

This Math Aftermath comes with a travel advisory of sorts for the potential explorer. In some sense—at least, in the author's sense—the material herein represents the mathematical zenith of the book: it's an extended journey into some other three-dimensional manifolds. While we wish to encourage the intrepid reader to forge ahead, we issue the advisory just in case you experience the Aftermath as an overwhelming deluge of math. If so, our advice is to jump to the next chapter until the feeling subsides. And with that, on to the math.

If we are willing to forego one-third of the Librarian's classic dictum that *the Library is a sphere whose exact center is any hexagon and whose circumference is unattainable* by yielding on the spherical nature of space, then there are two candidates for the large-scale shape of the Library, the *3-torus* and the *3-Klein bottle*, both worthy of our time and attention.[2] The two are intimately related, for the second can be thought of as the twisted, disoriented reassemblage of the first.

We'll proceed as we did earlier in the chapter: first, we'll gain an understanding of a two-dimensional object that lives in three dimensions, then we'll use that knowledge-base to visualize a three-dimensional manifold that lives in higher dimensions. This time, though, there will also be an intermediate step of reconfiguring our mind's eye to allow the hope of visualizing a *two*-dimensional object that lives most naturally in *four* dimensions. Finally, we'll briefly discuss the attributes of a Library modeled on either a 3-torus or 3-Klein bottle.

From the Plane to the Torus

We start with a familiar object, the everyday square, and then show that by "gluing" its edges together, various two-dimensional manifolds emerge. (Note that the square itself is NOT a manifold. Our rule is that it must be locally Euclidean, which we are taking to mean that if we stand at any point and take a few steps in any direction, we perceive ourselves as being in a Euclidean space. However, if we start at the edge of the square, we can't walk over the edge and still imagine ourselves in 2-space, for 2-space has no boundaries.)

Begin by marking the left and right sides with arrows pointing down, then continue by marking the top and bottom sides with double arrows pointing towards the right (figure 20). Now, identify the top and the bottom edges with each other, so that the arrows continue to point in the same direction. The mathematical sense of *identify* entails that the sides truly unify; it is as if they were never separate entities. By contrast, the best physical approximations are unfortunately coarse; one must glue, tape, or solder the edges together. Manifestly, after the mathematical identification the square has become a cylinder (figure 21).

Now identify the ends of the cylinder so that the arrows continue to revolve in the same direction—in 3-space, this is accomplished by bending the cylinder around so that the ends come together. When this identification is complete, the cylinder has transformed into a *torus*: the surface of a donut, the surface of a bagel—or, as topologists like to

FIGURE 20. A square with edges marked by orienting arrows.

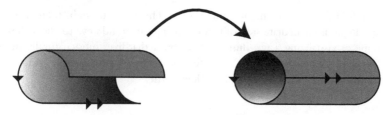

FIGURE 21. Identifying the top and bottom edges of the square.

point out, the surface of a coffee mug (figure 22).[3] (A statistician, Morris DeGroot, once jokingly remarked to me the literal truth that topologists don't know their asses from a hole in the ground.) This nifty sequence leads to the expression that *the torus is just a square with the edges identified to preserve orientation.* The torus is a 2-manifold; every point in it locally looks like the Euclidean plane. It has no boundary edges or walls, and if we think of it as a space into and of itself, like Euclidean space and the 3-sphere, the center is both everywhere and nowhere. The torus has an additional property which is quite extraordinary: it is *flat*, which means it can be embedded in Euclidean space in such a way that a bug walking between any two points on the torus could find a path whose distance is precisely the same as the straight-line distance between those two points on the square.

This should sound implausible; after all, the torus looks quite bent and the distance on the outer edge looks much longer than that on the inner edge. In fact, this is true; for the purposes of the illustrations and for boosting our intuition, we bent the cylinder until the ends met. We were purposely ambiguous and merely wrote "can be embedded in Euclidean space," several sentences back, rather than adding the key phrase: It must be *four-dimensional* Euclidean space. However, it's easy to see that the cylinder is truly flat in the geometric sense: Mark any two points on a

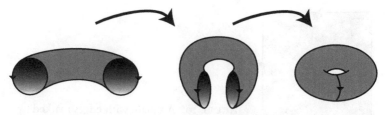

FIGURE 22. Identifying the left and right edges, forming a torus.

cylinder. Now let the cylinder unroll so that it is once again a square. Connect the dots in the square by a straight line. Now reroll the square into a cylinder. Voilà! Staying in the surface of the cylinder, the shortest distance between two points is, at most, the same as the distance between the two points in the unidentified square. (Why "at most"? Because there may well be a path crossing the identified edges of the cylinder that is even shorter than the straight-line path inherited from the square; regardless, by the unrolling/rerolling, we are guaranteed to achieve, at worst, the same distance on the torus as in the square.)

From the Plane to the Klein Bottle

The twisted, nonorientable reassemblage of the torus is called the *Klein bottle*. We form it by starting, once again, with a square. Again, mark double arrows on the top and bottom sides so that the arrows point in the same direction. This time, though, we place the arrows on the left and right sides so that they point in *opposite* directions (figure 23). Again, we identify the top and bottom sides with the arrows pointing in the same direction and thereby obtain a cylinder. This time, however, when we try to identify the ends of the cylinder, there is an insurmountable problem in 3-space. No matter how we twist or turn the cylinder around, there is no way to put the ends together so that the arrows are revolving in the same direction (figure 24). Although the picture looks bleak—impossible, in fact—the last twisted cylinder actually provides a ray of hope. If we rotate the orienting arrow counterclockwise around over the top of the bottom end of the cylinder, it's still pointing in the correct direction, and we obtain the mildly cheering picture shown in figure 25. Twisted around like this, one opening above the other, the orientations of the end-pieces match up: they are both counterclockwise.

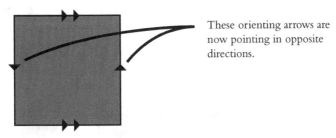

These orienting arrows are now pointing in opposite directions.

FIGURE 23. The orienting arrows for a Klein bottle.

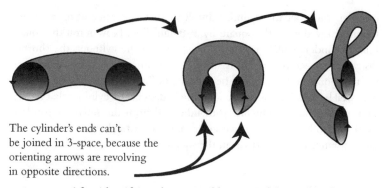

The cylinder's ends can't be joined in 3-space, because the orienting arrows are revolving in opposite directions.

FIGURE 24. After identifying the top and bottom of the square, forming a cylinder, problems ensue when trying to identify the left and right edges.

Because of the impossibility of aligning the cylinder ends, the Klein bottle cannot live in three dimensions; it requires at least four. A way of representing it in three dimensions is depicted in figure 26, but it requires a self-intersection. You can actually do this nicely by starting with a large enough piece of paper, marking the sides, taping the top and bottom edges together to make a cylinder, and then cutting a hole in the side to pass one end of the cylinder through. This is an excellent way to see how to allow the orienting arrows to point in the same direction. Perhaps this analogy will help explain why allowing the Klein bottle to be in four dimensions effaces the self-intersection. Suppose we confined ourselves to the two-dimensional Euclidean plane and were interested in joining a point inside a circle to a point outside the circle by a line (figure 27). Regardless of devious twists, turns, or serpentine path, it's pretty obvious that any curve joining the two points must intersect the circle somewhere.

FIGURE 25. After a counterclockwise slide, the bottom arrow remains consistent with its original orientation.

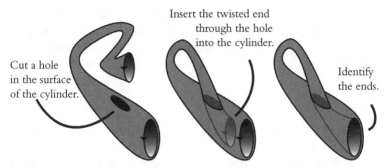

FIGURE 26. A two-dimensional depiction of a three-dimensional action that represents a 2-manifold which lives in four dimensions.

The only possible way to connect the points without intersecting the circle is to venture into the third dimension, pulling a path out of the plane (figure 28). Similarly, one may eliminate the self-intersection of the Klein bottle by simply pulling the offending part of the cylinder into the fourth dimension.

The "twisted" portion of the initial description of the Klein bottle comes from the fact that one could change the order of the construction by first identifying the left and right sides of the square before identifying the top and bottom. To identify the left and right sides, one must twist the square—and in so doing, create a Möbius band. If we did that, though, at this juncture it is very difficult to visualize how to glue the top and

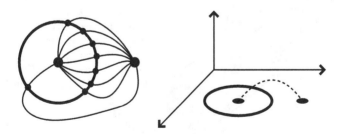

FIGURE 27 (left). In the plane, any line joining the two points must intersect the circle.

FIGURE 28 (right). Arcing into the third dimension allows the points to be joined without intersecting the circle.

bottom together to make the Klein bottle, because the top and bottom have been merged into one entity, a circle seemingly doubled on itself.

The "disoriented" portion stems from technical considerations and is manifest in two related, but distinct, ways. We'll cover them both below.

Outside Insights

Suppose we decide to walk a counterclockwise path on what appears to be the outside of the Klein bottle. In figure 29, the black arrow pointing out of the surface into space will represent our position as we start to walk, feet on the Klein bottle, head in the clouds.

Now some weighty philosophic problems naturally arise from even this innocent beginning. Euclid's plane and all 2-manifolds, including the Klein bottle, are "infinitely thin," much like the pages of the Book of Sand. Is the Euclidean plane therefore transparent? Does the plane, or any 2-manifold, possess a distinct "top" and "bottom"? (Borges makes playful use of these questions in his story "The Disk.") The mathematical perspective is that a path in the Euclidean plane or on a 2-manifold is

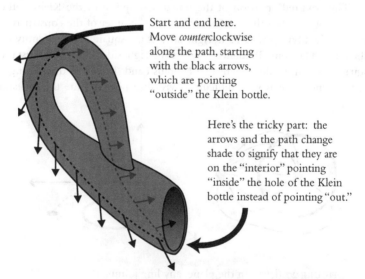

FIGURE 29. An arrow pointing "outwards" journeys around the Klein bottle and returns to the starting spot pointing "inwards." The path started "outside" the cylinder, which twists in four dimensions so the "outside" becomes the "inside."

simultaneously visible from both sides, and as such, it might be useful to imagine the Euclidean plane as a thin and supple sheet of transparent plastic. Then, any line painted on, for example, the top of the plastic is essentially visually indistinguishable from its image as seen through the plastic from below.

As we begin our walk along the surface, our feet naturally remain on the surface, while our heads naturally are "outside" in 3-space. Next, follow the path to where the two ends of the cylinder are identified. (This looks like the "hole" at the front of the Klein bottle in figure 26.) Note that as we enter the "hole", the arrow and the path both are faded to suggest that we are now inside the Klein bottle, and that our heads are now pointing "inside" rather than "outside." Keep moving "inside" the Klein bottle through the self-intersection—which isn't really there—until we've circled around to our starting point (figure 29). The arrow representing us was initially pointing "out" and now it is pointing "in." The Klein bottle, which has neither holes nor boundaries, also has no inside or outside in the sense that we intuitively understand these terms—a disorienting revelation indeed.

A faintly analogous situation occurs with the familiar circle. In the plane, there's a distinct inside and outside—look again at figure 27. As discussed earlier in the chapter, in 3-space the circle has nothing easily definable as an "inside" or an "outside." It certainly does not cut 3-space into two eternally separate pieces, as does, for example, a 2-sphere. The correspondence between the Klein bottle's and the circle's lack of an inside and an outside hinges on dimensionality. A circle is a one-dimensional object that can live in two dimensions. If the circle is in the plane, in 2-space, then the "dimensional difference"—technically, the *codimension*—is equal to one:

$$2 - 1 = \text{dimension(2-space)} - \text{dimension(circle)} = 1.$$

On the other hand, if the circle is in 3-space, the codimension is equal to two:

$$3 - 1 = \text{dimension(3-space)} - \text{dimension(circle)} = 2.$$

Similarly, if the Klein bottle, a two-dimensional object, is in 4-space, the codimension is once again equal to two. A codimension greater than one implies that the object can't separate the space into two distinct pieces; thus, there can be no inside or outside.

Summarizing, the Klein bottle is an example of a *one-sided* 2-manifold with no boundary. (By contrast, the Möbius band, another delightfully disorienting object, has a boundary, an edge.) All the boundaryless 2-manifolds familiar to us from our sensual life in 3-space have an inside and an outside—think of a sphere, a torus, the surface of a pretzel, or the surface of any familiar object. They all cut space into two distinct pieces.

The Klein bottle does not separate space—it has one side only, and there is no way of distinguishing between the inside and the outside. Moreover, in four-dimensional Euclidean space, the Klein bottle is geometrically flat for the same reasons as the torus: pick any two points on it, reverse the identifications back to a square, then draw the straight line that connects the two points. An unimaginable construct, to say the least.

Inside Outsights

There is another disorientation involving the Klein bottle. For this one, we imagine, taking a cue from *Flatland*, that we live a two-dimensional existence *wholly contained within the surface* of the Klein bottle. Outside and inside are meaningless words to us: the Klein bottle is our entire universe. Befitting our new planar existence, let us take a new form, that of a flag rather than an arrow. The flag that we are curves and bends with the Klein bottle as we move around; again, it is—we are—wholly contained within the universe that is the Klein bottle.

Again, this time the black flag is part of the surface of the Klein bottle, not perpendicular to the surface like the arrows in the previous illustration. See figure 30. Now move the flag counterclockwise along a path that exploits the one-sided nature of the Klein bottle, the same path as in the previous section. As in figure 29, we change the shade of both the path and the flag as they proceed from the "outside" of the Klein bottle through the "hole" to the "inside." Note that throughout our journey, the black flag points in the direction of motion.

Observe that although the flag begins its journey with the pole pointing in one direction on the surface of the Klein bottle, after it has slid around to the starting point the pole, still contained in the universe that is the Klein bottle, is now pointing in the other

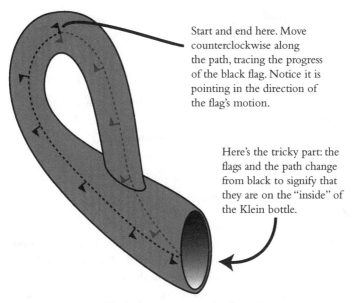

FIGURE 30. A black flag returns as if reflected in a mirror.

direction. Perhaps, jaded by the one-sided oddity of the Klein bottle, this isn't a big surprise—after all, it is easy to imagine slithering around on the floor and ending up with our feet located at their initial spot and with our head pointed in the opposite direction than at the start.

More disorienting, though, *the black flag has come back a mirror reflection of itself.* There is no obvious intuitive analogue for us. Any journey you take, transformative though it may be, will not result in your coming back as a mirror-reflected image of yourself. You may, for example, feel a shadow of your former self, or half the person you used to be, or find your partner besieged by 50 suitors; regardless, it will not be the case that your heart is now, from everyone else's perspective, on the right-hand side of your body. Figure 31 illustrates the categorical difference between a rotation and a mirror-reflection.

On a sphere, on a torus, or in the Euclidean plane, any journey the flag might take would result in, at worst, a rotation. There is no possible path that allows for the flag to be mirror-reflected—a journey into the fourth spatial dimension is required.

Rotating the black flag in the plane

Mirror reflection of the black flag

FIGURE 31. The flag cannot be spun to produce a mirror-reflection.

From 2-Manifolds to 3-Manifolds

Let's generalize these 2-manifolds, the torus and the Klein bottle, to their three-dimensional equivalents, the 3-torus and the 3-Klein bottle. To do so, we start with a solid cube, instead of a square, and identify opposite sides, and mimicking what we did in two dimensions, we'll begin by creating a 3-torus. Again, we'll take advantage of working in three dimensions to bend the cube to identify the sides; the natural space for the identifications is *six*-dimensional Euclidean space: in six dimensions, the 3-torus is flat. If we are willing to have a curved, distorted representation akin to the 2-torus in three dimensions, a "mere" four Euclidean dimensions suffices to hold the 3-torus. For the 3-torus, arrows are insufficient to specify an orientation of a face of the cube, but spirals will serve. (Think about why this should be so.)

Figure 32 shows the initial solid block inscribed with appropriate spirals. This time, we bend the sides of the cube around, identifying the left and right faces while taking care that the spirals being glued

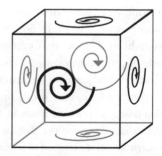

FIGURE 32. A solid cube, with the faces oriented by spirals.

TOPOLOGY AND COSMOLOGY   81

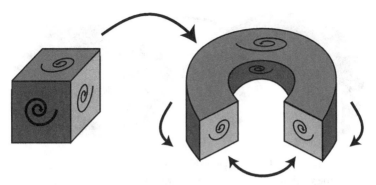

FIGURE 33. Bending the cube in 3-space to identify same-colored left and right sides so that the spirals' orientations align correctly.

together spin in harmony (figure 33). After this first identification, we are confronted with a millstone, whose top and bottom must be identified. (The inside and outside are, of course, also identified. We'll discuss that afterwards.) Turn the millstone sideways—and shrink it—to make it easier to visualize this step (figure 34). Proceed by identifying the visible gray ring on the right-hand side of the millstone with the hidden gray ring on the left-hand side (figure 35).

Now, we are presented with a donut that has a smaller donut drilled out of the middle of its interior; a donut waiting for a filling, as it were. A donut with a non-donut inside. The surface of the exterior donut is a 2-torus, while the surface of the interior non-donut is also a 2-torus. These two tori correspond to the front and back square faces of the initial

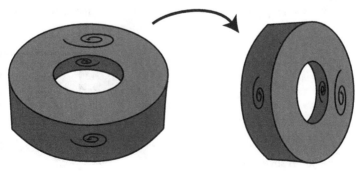

FIGURE 34. Turning the millstone on its side.

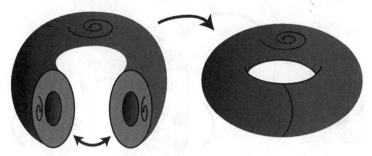

FIGURE 35. Step II: Identifying the cube's top and bottom (the millstone's left and right sides) so that the orienting spirals align.

cube, and they sprang into being when, in the process of identifying the other four faces of the cube, *only the edges of these squares were identified*. Now, the (invisible) interior 2-torus must be identified with the (visible) exterior 2-torus. By tugging them both into the fourth dimension, where they no longer divide space into an "inside" and "outside," they may be glued together, producing the 3-torus.

Before moving on, let's look at one more way to visualize a 3-torus. Once again, we'll proceed by analogy with the eminently imaginable 2-torus. If we take a 2-torus and intersect it with a plane (as in figure 36), the result is a circle (figure 36). Another way to see this is to take a slice of the square that becomes the 2-torus (figure 37). If we think of a 2-torus in this way and flatten it out onto the plane, we may represent it as a circle of circles (figure 38).

If we take a three-dimensional slice of a 3-torus, we get a 2-torus. One way to see that is to look at a slice of the solid cube we started with (figure 39). Consequently, each slice of the 3-torus is "flattened" out into

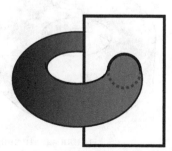

FIGURE 36. A slice of one side of the 2-torus yields a circle.

TOPOLOGY AND COSMOLOGY 83

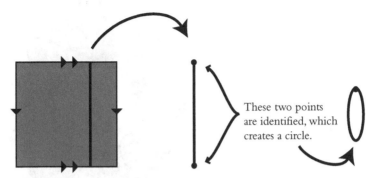

FIGURE 37. A line with identified endpoints is a circle.

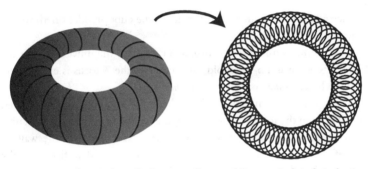

FIGURE 38. The circles of a 2-torus, flattened into a circle of circles in the plane.

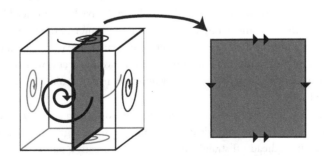

FIGURE 39. A slice of the 3-torus is a square with identified edges—a torus.

FIGURE 40. The circle of 2-tori, "flattened" out into 3-space.

3-space as a 2-torus. Since the two sides of the cube are identified, we get a circle of 2-tori (figure 40).

Let us now consider the 3-torus as a model for the universe that is the Library. Since it is a 3-manifold, the center of the 3-torus is everywhere and nowhere, so *the exact center is any hexagon*.

Next, there is a sense in which the 3-torus has sorts of circumferences, which arise in the following ways. Imagine we're at the center of the cube, facing "out of the page." If we move to the exact center of the wall on our left, when we reach it, due to the fact that it is identified with the right-hand wall, the "left wall" is simultaneously the "right wall," which is actually no wall, but rather an unrestricted passage back to the other side of the initial cube. So if we continue to move, we'll end up back where we started. (For that matter, if it is a small 3-torus, if we turn our head and look to either left or right, we'll see the back of our head.)

Similarly, if we moved up or down from the center of the initial cube, we'd again end up back at the center of the cube. Finally, if we moved forward or backwards, the same phenomenon would occur, which means that in a small 3-torus, looking in *any* direction means looking at the back of our head.

In a 3-sphere, if we head off straight in any direction and stay straight, we'll eventually circumnavigate the sphere along a great circle. In a 3-torus, if we head off straight in a particular direction and stay straight, depending on the angle we set out we will eventually either end up exactly where we started or else come arbitrarily close to our initial point. If the Library is a 3-torus, by dint of its enormity, again *all of its circumferences are unattainable* by a librarian. Moreover, since *there are no boundaries* to it, the 3-torus is *limitless*. Because journeying straight in any direction would eventually return us to where we began, the 3-torus is *periodic*. Therefore the 3-torus satisfies two of the three conditions of the classic dictum and all three of the conditions of the librarian's solution; the only condition is misses is that it isn't a sphere.

It is also geometrically flat, in the same sense as the 2-torus, which might be a desirable quality for the Library. Although a large enough sphere, such as the earth, will appear flat, a sphere is always curved. Considering the Library as a 3-torus embedded in 6-space, there'd be absolutely no way, locally, for the librarians to determine that they are living in a 3-torus as opposed to living in Euclidean 3-space.

This leads to some highly speculative questions. What if the hypothetical Builder(s) of the Library wished to test the librarians? If the Library was a 3-sphere and the librarians grew tremendously technologically advanced—more than us—they might develop a method to measure the local curvature of space. If they discovered that the curvature was nonzero, they'd know that the librarian's solution at the end of the story was false: Euclidean 3-space has no curvature. If, on the other hand, they found the curvature to be zero, they would have to face the bitter realization that once again, they didn't possess enough information to decode the topology of the Library.

The Library includes mirrors. Borges draws our attention to this via the following passage, part of the description of the particulars of the makeup of the Library:

> In the vestibule there is a mirror, which faithfully duplicates appearances. Men often infer from this mirror that the Library is not infinite—if it were, what need would there be for that illusory replication? I prefer to dream that burnished surfaces are a figuration and promise of the infinite....

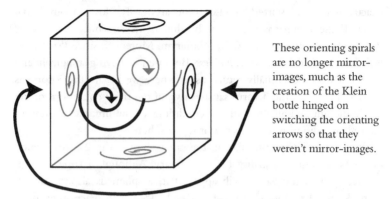

These orienting spirals are no longer mirror-images, much as the creation of the Klein bottle hinged on switching the orienting arrows so that they weren't mirror-images.

FIGURE 41. A solid cube, with one different orienting spiral than in figure 32.

After the development of our final 3-manifold, we'll submit a fanciful explanation accounting for the presence of the mirrors.

Begin by reversing the spin of one of the orienting spirals, and next identify the opposite faces of the initial cube as we did creating the 3-torus (figure 41). The outcome will be a three-dimensional Klein bottle, which we'll call the *3-Klein bottle*. As with the 3-torus, we first endeavor to identify the left and right faces of the solid cube; this time, though, we are unable to accomplish the first step in three dimensions. Look closely at the left-hand "bent-square" in figure 42. The spiral on both the left-hand square and the right-hand square are turning clockwise. Thus, if we naively try to put the two squares together as we did in creating the 3-torus, the orientations do not align. Rotating either of the squares will not affect this problem, as the mere fact of the rotation will not impinge upon the spiral's clockwise orientation. (Think of it this way: imagine walking up to your reflection in the mirror and attempting to touch your right hand with its reflection. Easy to do. However, if your identical twin walked up to you and you both held out your right hands in the same fashion, your hands wouldn't align or touch. This is why the spirals need to be mirror-reflected, that is, flowing in opposite directions, for the sides to identify.)

As with the Klein bottle, bending and twisting the cube up and around allows the spirals to be in the same alignment when placed one over the

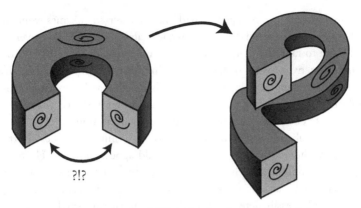

FIGURE 42. Similar to the 2-Klein bottle, these faces can't be joined—both faces have clockwise spirals on them. Since they must be glued "face-to-face," the spirals need to be mirror-reflected orientation to properly align. Compare with figure 24.

other. Again, as with the Klein bottle, the oriented squares cannot be joined in 3-space. To do so, we must again bend part of the cube "up" into the fourth dimension, precisely the same as we did with the 2-Klein bottle. (Unfortunately, due to the solidity of the interior of the cube, this is beyond our ability to effectively illustrate: rather than a simple circle of self-intersection, we'd be confronted with a truncated solid pyramid of self-intersection contained in the interior of the original cube. The top square of the solid truncated pyramid would be where the top light-gray square "entered" the original cube, and the bottom of the solid truncated pyramid would be the joined pair of light-gray squares facing us in the front.) And then we must still perform other identifications!

The 3-Klein bottle can also be embedded so that it is flat; furthermore, it enjoys many of the other properties of the 3-torus as well. It is therefore a reasonable candidate for the topology of the Library.

However, if an intrepid nomadic civilization of librarians or a band of immortal librarians managed to walk a loop that took them through the identified disorienting faces, they would find that they would appear normal to themselves, but when they returned to where they began, the Library would be seen as if reflected in a mirror. The Library wouldn't have changed; rather, it is the librarians' perspectives that would have been turned inside out—in fact, it's an interesting question whether or not such mirror-reversed people with mirror-reversed enzymes would be able to eat our food and digest it to extract nutritional value. If we were to ask them to raise their right hand, they would raise their left hand (from our perspective), while truthfully swearing (from their perspective) that they were raising their right hand. This is exactly parallel to the mirror-reflection of the black flag in figure 30 and in figure 31 that occurs after a complete circuit through the disorienting identification.

If the Library appeared as reflected in a mirror to the inverted librarians, there are some things that would appear different. However, by making only a few changes to the structure of the Library, we can disorient the librarians so that if they should manage to make such a loop, they wouldn't easily detect that they've been mirror-inverted.

The first problem revolves around the spiral staircases. They might all be subject to a rule such as "walking clockwise means going down" (figure 43). When the librarians cross through the disorienting face, they will find that the rule has become "walking clockwise means going up." The easy way to remedy this staircase asymmetry is to "insist" the builders of the Library randomly designate different spiral staircases to go up or down when traversed clockwise. Similarly, the sleeping compartments, the lavatory closets, and the mirrors must be randomly distributed on left and right sides of the entrances.

Another, and perhaps the most important, visual asymmetry is that the orthographic symbols will be mirror-reversed. For an example, see

FIGURE 43. On the left-hand spiral staircase, walking clockwise means walking down. On the right, walking counterclockwise takes a librarian down.

FROWN   versus   ИWOЯᖴ

FIGURE 44. The word "FROWN," reflected in a mirror.

figure 44. An elegant way to avoid this asymmetry is to specify an alphabet whose orthographic symbols are invariant under left-right flips; typically, this is called *bilateral symmetry*. Here are 25 invariant Roman letters and symbols from a standard computer keyboard.

A H I M O T U V W X Y 8 ' " —

= + . : * ^ | ! . (blank space)

(A sharp-eyed reader will note that some of the letters in this font, such as A, M, U, V, W, X, and Y, aren't precisely bilaterally symmetric. These letters need only minor modifications to become bilaterally symmetric.) Fourteen other symbols, readily available, are also invariant under mirror-inversions:

_ † ° ∞ ± ∏ Ω ¡ ∆ ÷ ◇ ‡ Â ˘

Furthermore, there are pairs of symbols that when flipped produce each other:

( ) [ ] { } < >

It wouldn't take long to create an aesthetically pleasing set of 25 symbols with the desired mirror-reversal invariance.

Along similar lines, almost all book titles are printed on the spine so that if the book is laying flat on a table and the front cover is visible, the title can be read: the *tops* of the letters abut the *front* cover of the book. Let's call this "top-front" labeling. Try vertically holding the spine of a top-front book to a mirror. Not only are the letters mirror-reversed, but the title now appears as a "top-back" label on the spine (figure 45). A solution to this problem is to simply write the titles of the books vertically down the spines (figure 46). This way, even with a mirror-reversal, a librarian wouldn't notice anything amiss.[4]

Moreover, it's ironic that a band of immortal librarians who circumnavigated the 3-Klein bottle Library and returned to their originating

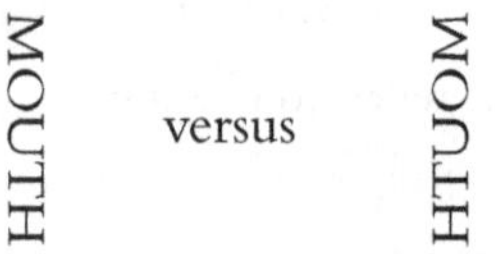

MOUTH versus HTUOM	M O U T H versus	M O U T H

FIGURE 45. "MOUTH," written horizontally on the spine, reflected in a mirror.

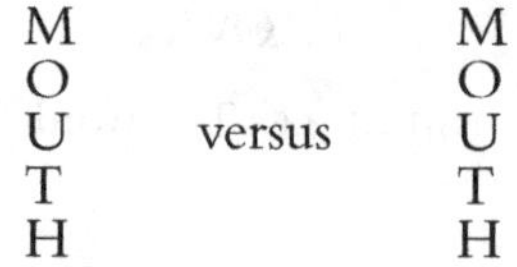

FIGURE 46. "MOUTH," written vertically down the spine, reflected in a mirror.

hexagon *wouldn't recognize any of the books*. Although the titles would all be the same, the contents would all look different. Open a book to the first page while looking in a mirror: it appears that the book is open to the *last* page, not the first. So for the books in a particular hexagon to read the same to a mirror-reversed librarian, the hexagon would need to consist of books that were 410-page palindromes! However, if they were intrepid enough to complete a second circumambulation of the Library, they'd experience a mirror-reversal a second time and then everything would look the same as when they started out.

Suppose that the constructors of the Library incorporated these design changes to the physical structure and the orthographic symbols. If all the librarians migrated, the Library would not look mirror-reversed to them, even after passing through the disorienting identified faces. However, if they split into two groups and one group managed to circumnavigate the Library, the descendants of the nomadic group would return and discover—from their perspective—a foreign group of librarians who didn't know left from right. Although it's more likely that any such disparity would be attributed to language differences, a librarian of genius might realize the significance of the invariance of the orthographic symbols when reflected in the mirrors. Such a mathematically minded librarian might then deduce that the Library is a nonorientable 3-manifold, and a 3-Klein bottle would surface as the most likely candidate. Such a librarian would know more about the topology of the Library than we know about our own universe.

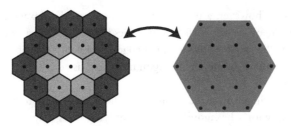

FIGURE 47. Consider rings of adjacency as defining hexagons on each floor.

We've now covered all the hard parts for the last topology we wish to propose for the Library. The end result will once again be either a 3-torus or a 3-Klein bottle; we will just take a different, more elegant, path to achieve it. In our discussions of the tori and Klein bottles, we began with a square or a solid cube and then, respectively, identified edges or faces.[5]

In the next chapter, we'll look at a single floor of the Library, in part by choosing an initial hexagon and considering rings of adjacencies to it. Any ring of adjacency, combined with the hexagons contained inside, forms another shape, which is essentially hexagonal in nature (figure 47). (This is particularly clear if we look at the midpoints of the hexagons.) If we start with a hexagon and carefully identify the sides, just as when we start with a square, the resulting object is either a torus or a Klein bottle. Jeff Weeks, in *The Shape of Space*, pages 116–26, discusses this and provides very clear illustrations, and Weeks' detailed explanation of these issues is both elegant and relatively accessible. For a reader wanting to know more about 2- and 3-manifolds, it is an excellent reference.

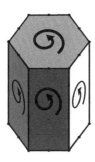

FIGURE 48. A hexagonal prism whose same-color, spiral-oriented faces will be identified. (The other four sides aren't shown.)

There is, however, a major difference between the hexagon and the square as the object that will have its edges identified. It turns out that to embed the "hexagon with identified edges" in Euclidean 3-space, which is *not* its "natural" home, the hexagon must be twisted as well as stretched and bent.

Generalizing from two dimensions, it shouldn't be hard to believe that by starting with a hexagonal prism, identifying faces may yield either a 3-torus or a 3-Klein bottle (figure 48). This model of the Library has the additionally pleasing aspect that, in some difficult-to-define sense, the larger geometric structure mimics the smaller structure of the hexagon. We wrote "difficult-to-define" because as soon as the faces are identified, the hexagonal prism is subsumed into the 3-torus or 3-Klein bottle. Again, a torus may be considered a square with its edges identified. However, once the edges are identified, the edges are gone and the square is gone: all that's left is the torus. The identified edges may be drawn in for the purpose of clarifying the process, but they are no longer there. Equally possible, a torus may be a hexagon with its edges identified. In both cases, the resulting object is a torus, but the differing characters of the square and hexagon may leave a detectable, classifiable trace.

A Library modeled on the 3-torus or 3-Klein bottle could be based on either a cube or a hexagonal prism with identified faces. It's conceivable that an immortal librarian of genius, endowed with a means of measuring curvature and possessed of an infinite photographic memory, who repeatedly traversed the Library might also some day be able to guess the Library's topologic structure. Regardless, we who may now consider ourselves the architects of the Library may feel the special glow that derives from the outlining of exquisite solutions to a demanding problem.

FIVE

Geometry and Graph Theory
Ambiguity and Access

> *If one does not expect the unexpected one will not find it out.*
> —Heraclitus, Fragment 18

> *A library is a collection of possible futures.*
> —John Barth, *Further Fridays*

THE LIBRARY, AS EVOKED IN THE STORY, HAS inspired many artists and architects to provide a graphic or atmospheric rendition of the interior. These range from Stefano Imbert's lean and elegant drawing adorning the cover of this book, to the deliberately alienated Piranesi-like drawings of Desmazières in the Godine Press edition of *The Library of Babel*, to Toca's beautifully symmetric honeycombs in *Architecture and Urbanism*, to Packer's bold expressionist frontispiece in the Folio edition of *Labyrinths*, to a host of illustrations easily found online. All of these illustrations sacrifice, to some degree, accuracy in favor of artistic effect. For example, even the cover illustration of this book locates the spiral staircase in the center of the hexagon, whereas Borges writes (emphases added):

> The universe (which others call the Library) is composed of an indefinite, perhaps infinite number of hexagonal galleries. In the center of each gallery is a ventilation shaft, bounded by a low railing. From any hexagon one can see the floors above and below—one after another, endlessly. The arrangement of the galleries is always the same: Twenty bookshelves, five to each side, line four of the hexagon's six sides; their height of the

bookshelves, floor to ceiling, is hardly greater than the height of a normal librarian. *One of the hexagon's free sides opens onto a narrow sort of vestibule, which in turn opens onto another gallery, identical to the first—identical in fact to all.* To the left and the right of the vestibule are two tiny compartments. One is for sleeping, upright; the other, for satisfying one's physical necessities. *Through this space, too, there passes a spiral staircase,* which winds upward and downward into the remotest distance.

As we shall see, quite a lot pivots on the ambiguity arising from the italicized phrase "One of the hexagon's free sides opens onto a narrow sort of vestibule, which in turn opens onto another gallery, identical to the first—identical in fact to all." (This passage is equally uncertain in Spanish, "Una de las caras libres da a un angosto zaguán, que desemboca en otra galería, idéntica a la primera y a todas.")

Now, each hexagon has six sides, two sides of which lead to another hexagon. If Borges meant that *each* one of the free sides gives upon a narrow entrance way with two miniature rooms, then it follows that in *every* doorway, there is a spiral staircase, rising and sinking beyond sight. Surprisingly, profound and prodigious consequences derive from this doubled staircase arrangement.

On the other hand, we may read the italicized phrase in a different way. If Borges meant that *exactly one* of the free sides gives upon a narrow entrance way with two miniature rooms, then it follows that the Library contains *pairs* of hexagons, joined by small rooms and spiral staircases. Although the difference may seem slight, in this variation of the Library, not only are the plumbing and spiral staircase construction costs cut in half, it actually turns out to be the case that the librarians can lead very different kinds of lives than in the first scenario.

Before we illustrate these two possibilities, in the service of verisimilitude let us sensibly estimate the dimensions of one hexagon and then sketch it. In his "Autobiographical Essay," appearing in *The Aleph and Other Stories*, pages 243–44, Borges notes that

> My Kafkian story "The Library of Babel" was meant as a nightmare version or amplification of that municipal library [the Miguel Cané Municipal Library], and certain details in the text have no particular meaning. Clever critics have worried over those ciphers and generously endowed them with mystic significance.

We were fortunate to be able to visit the Miguel Cané Municipal Library in Buenos Aires, and also both the old and new National Libraries of Argentina. The first three measurements below come from the Miguel Cané Municipal Library, while the fourth comes from a narrow and steep marble spiral staircase in the old National Library.

> Length of bookshelf: 3 meters (large double-sided bookcase)
> Depth of bookshelf: 0.3 m
> Height of bookcase: $\sim$ 2.21 m
> Diameter of spiral staircase: $\sim$ 1 m
>
> Miniature room for standing sleeping: $\sim$ 0.5 m by 0.5 m
> Miniature room for relief of physical necessities: $\sim$ 0.5 m by 0.5 m
> Walking space between staircase and walls: $\sim$ 0.5 m

Thus the approximate length of each hexagon's side needs to be three meters, which corresponds nicely to the actual size of the original bookcases at the Miguel Cané Municipal Library. Based on the size of the (presumably square) miniature rooms, the thickness of the walls of each hexagon should be approximately 0.5 meters, although it appears the builders could get by with walls 0.25 meters thick. See figure 49 for the layout of a hexagon.

Another pertinent item is what Beatriz Sarlo notes on page 71 of *Jorge Luis Borges: A Writer on the Edge*, "As Borges himself declared in an

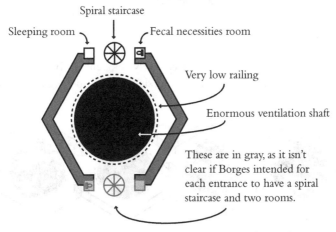

FIGURE 49. A diagram of a typical hexagon of the Library.

96   UNIMAGINABLE MATHEMATICS

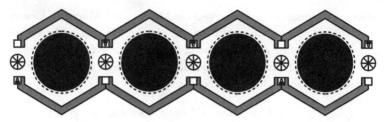

FIGURE 50. Hexagons with a spiral staircase in every doorway.

interview, his first spatial idea for the Library of Babel was to describe it as an infinite combination of circles, but he was annoyed with the idea that the circles, when put in a total structure, would have vacant spaces." From the description in the story and especially this quote from Borges, we conclude, as have a number of other commentators, that the Library resembles a honeycomb with no interstices.

Figures 50 and 51 are a pair of illustrations putting together the hexagons; for the first, we show four conjoined hexagons modeled with a spiral staircase appearing in every doorway (figure 50). Based on the second interpretation, figure 51 shows a linear arrangement of hexagons designed with the staircases and two small rooms located in every other doorway.

Now, we'll make three general observations. Then we'll first assume that each and every doorway has a spiral staircase and see what consequences ensue. After that, we'll examine what happens when alternating doorways are pierced by a spiral staircase.

Three General Observations

The most important fact is deceptively hidden in Borges' simple phrasing, "Twenty shelves—five long shelves per side—cover all sides

FIGURE 51. Hexagons with a spiral staircase in every other doorway.

GEOMETRY AND GRAPH THEORY 97

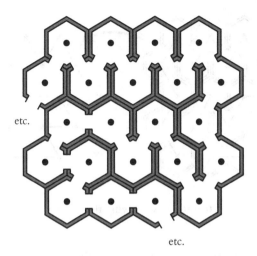

FIGURE 52. A contorted passage through the Library.

except two..." This obviously means that each hexagon has two doors, but when combined with the snug nesting of the honeycombed hexagons, it means that a librarian disdaining the stairs may only move forward into a new hexagon or double back to the previously visited hexagon. As in a labyrinth, a librarian remaining on the same floor has one path only to tread.

The next observation is that it's conceivable the floor plan of a level of the Library may look like the preceding illustrations, in which the paths run straight through the hexagons. However, it is consistent with the text—and the atmosphere of the story—that the corridors weave and spiral around symmetrically or chaotically (figure 52). (In the succeeding pictures, in the interests of graphic clarity, we omit the miniature rooms, the very low fence, and the spiral staircases, and shrink the enormous ventilation shafts to small black dots.)

The last of the three observations is practical rather than structural or theoretical. It also provides a nice example of "thinking like a mathematician." If we use the shovel, pick, and whisk of our analytical imagination to pare away the obscuring accretions of reality, we reveal the artifacts of our ideas, which provide the wherewithal to build a theory. In this case, we collapse each hexagon to a point, represent the passageways by lines connecting the dots, and throw away the walls and bookcases of the hexagons (figure 53). With these notions and simplifications in hand, let us journey to the first of the two Libraries.

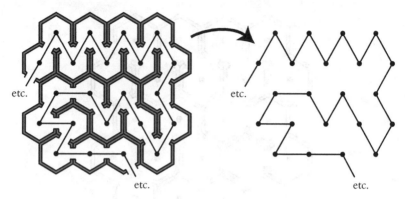

FIGURE 53. Clarifying the Library.

In Every Doorway, There Is a Spiral Staircase

Assume that every doorway is intersected by a spiral staircase, regardless of pattern of floor plan. This prevalence of spiral staircases led to an inkling, then a hunch that became a surmise, which we ultimately formulated as a conjecture and subsequently proved. We approach this theorem as a variation of the locked-room detective story, and hope that H. Bustos Domecq, Borges and Bioy Casares' fictional anti-detective, would admire it.

We are librarians talking in a hexagon about the significance of the 25 orthographic symbols that comprise the markings in the books when, from an adjacent hexagon, we hear raised voices shouting muffled words that are difficult to comprehend; only the rage is clear. We hear thuds, now, as the violence escalates. We look at each other, shocked, and peer through the doorways into two of the six hexagons adjacent to ours. As far as we can see through the portals, the nearby hexagons are empty. Without any discussion, acting on impulses born of common humanity, we each dart through one of the doorways leading out of our hexagon. As we both scan the exit passages of the hexagons we've just run into, the same thought, remarkably, simultaneously enters our minds:

> *Will I, or my friend, necessarily be able to reach the adjacent hexagon in time to prevent a murder?*

Continuing to run into connecting hexagons through the unique entrance doorway and running out through the unique exit passage, we each assure ourselves that the sounds truly came through the wall from a hexagon abutting the one in which we were talking. (For example, the sounds did not float down the airshaft or up a spiral staircase.)

After running until exhausted, you perceive an omnipresent and ominous silence overwhelming the intermittent gasps of your fragile breathing. Defeated, trembling, you reverse direction: your exit doorways become entrance passages and vice versa. There is no chance that you will become lost. The hope you cherish, that which gives you strength enough to trudge back to the starting-point hexagonal gallery, is that I was able to reach the adjacent hexagon in time to temper the dispute.

My crestfallen look, lit only by my eyes eagerly seeking to read your face, is enough to tell you: I failed, also. We sit uncomfortably by the air shaft and each recount the particulars of our fruitless runs; there's not much to say, "I ran in, I ran out; I ran in, I ran out; I ran in, I ran out; I ran in, I ran out; ..."

It is irrelevant which one of us first exclaimed, "The stairs! The spiral stairs! Maybe we could have gone up, over, and down to the hexagon. Or down, over, and up? Or down, down, over, up, over, up, up, over, and back down?" It is equally irrelevant which combination of whose thoughts destroyed this diaphanous, infinitely permutable, insight:

Every hexagon has an airshaft through the center—it is easy enough to look up and down our airshaft and see what we always see: stacked hexagons. Moreover, each hexagon has two sides singled out by the presence of a spiral staircase (figure 54). Looking up and down the spiral staircases of our hexagon confirms what we already know, that the hexagons above and below must have the staircases in the same sides as the one we are in. In other words, the hexagons above and below are, in this regard, exact clones of our hexagon. Extending this reasoning, for each and every hexagon on our floor, the hexagons above and below it are exact clones. That means that the labyrinthine paths we ran are precisely the same above and below—the stairs and shafts dictate this.

> Each floor plan is inevitably, invariably, precisely the same as every other floor plan. There is no advantage gained by taking a set of stairs up or down.

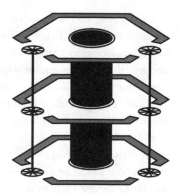

FIGURE 54. The spiral staircases and the air shaft dictate the geometry.

We may, therefore, restrict our investigation to the floor we are on. We might have been lucky; it could have been the case that our starting-point hexagon comprised a part of the floor plan that looked like figure 55. In such a case, one of us would have reached the hexagon in seconds. On the other hand, we were unlucky, so unlucky that we couldn't say how unlucky we were (figure 56). We didn't hazard a guess as to how many hexagons we would have to pass through to

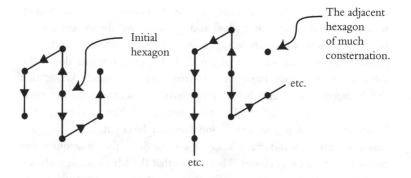

FIGURE 55 (left). A path that quickly visits all six adjacent vertices.

FIGURE 56 (right). A path that traverses an unknown number of hexagons before reaching the hexagon adjacent to our starting point.

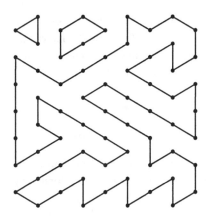

FIGURE 57. Part of a floor plan that includes three closed loops.

reach the other librarians; such a presumptive act would be tantamount to heresy. For that matter, after thinking about it, we daren't even say if the other librarians were really fighting; perhaps they belonged to an entirely different civilization—perhaps an entirely different *species*.

Our Stark and Depressing Realization: *Our question was decisively answered. We would not necessarily be able to reach the adjacent hexagon in time to prevent a murder.*

Our conception of the Library's structure was so perturbed by these cascades of devastating insights that it didn't even occur to us until later that the floor plan could plausibly contain eternally inaccessible closed loops, such as the three in figure 57.

The unexpected, against our desires, found us and found us wanting. Without the barrier of even a single door, the adjacent hexagon—the source of noise, confusion, and probable violence—was locked away from us forever.

(We were very startled to realize this; indeed, it only became clear while flying to Buenos Aires when we sketched out the library floor plan. Our readings of "The Library of Babel" always left us with the impression that regardless of which hexagon "we" were "in," we could reach any nearby hexagon in a short period of time. The Realization of this section, in fact, a minor *lemma* in the field of graph theory, is a prime example of the unimagined math of the story. The Math Aftermath, "A Labyrinth, Not a Maze," at the end of this chapter contains an extension of this story providing a sense of why a stronger result about the inaccessibility of adjacent hexagons must be true.)

There Is One Spiral Staircase Per Pair of Hexagons

Now let's examine the Library by interpreting the first paragraph of the story to mean that only one passageway in each hexagon is perforated by a spiral staircase (as in figure 51). Once again, the hexagons are forced to be stacked by the existence of the ventilation shafts, but in this case, only one wall, one doorway, is specified by the spiral staircase. This entails that although the hexagons' sides are aligned, they are no longer cloned. Hexagons above and below each other must share one entranceway, but the second may branch out in a different direction (as in figure 58).

Thus, a pair of floors may have labyrinth patterns such as these depicted in figure 59. These illustrations include the spiral staircases, because the presence of a staircase induces a connecting passage between hexagons on all other floors of the Library. Combining the floor plans produces the pleasingly symmetric picture of figure 60. Most importantly, this means that a librarian may reach any adjacent hexagon by traveling through only two additional hexagons and two flights of stairs, one up, one down. The sense of a bewildering array of choices is omnipresent and factual.

It is our considered opinion that Borges simultaneously intended for the Library to have a spiral staircase in every doorway and also to present the librarians with a bewildering array of options. The "stark and depressing realization" of the librarians in the preceding section indicates the impossibility of such a conjunction, whereas this minor modification allows for enormous mutability in the floor plans and potentially a quick access to any nearby hexagon.

GEOMETRY AND GRAPH THEORY   103

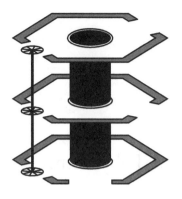

FIGURE 58. Note the three different doorways to the heterogenous hexagons.

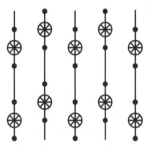 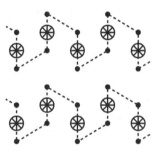

FIGURE 59. Two different floor plans allowed in this version of the Library.

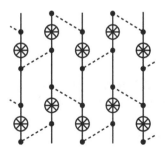

FIGURE 60. An overlay of the two floor plans, which illustrates the complete accessibility of all adjacent hexagons.

Math Aftermath: A Labyrinth, Not a Maze

> *The subject does not belong to the world; rather, it is a limit of the world.*
> —Ludwig Wittgenstein, *Tractatus Logico-Philosophicus*

The goal of this Math Aftermath is to provide grounds for believing an even stronger consequence than the "stark and depressing conclusion" that followed from the first case we deconstructed, that adjacent hexagons need not be accessible. The stronger consequence has a relatively easy proof, but is too messy to offer up in these pages due to the necessity of breaking down a number of related cases. (Despite not offering much of a framework for the proof, it is tempting to employ the standard trope appearing in works of mathematics at moments such as this: "The reader may supply the details.") The librarians in this pastiche will name it "our conjecture of extreme disconsolation," and it is:

> In any Library constrained as this one,
> on any given floor,
> for any positive integer n less than, say, $1{,}000{,}000{,}000{,}000 = 10^{12}$,
> there must necessarily be pairs of abutting hexagons, H_1 and H_2,
> such that a librarian would need to walk through more than n
>   distinct hexagons to travel from H_1 to H_2.

That is, there are many, many hexagons that possess effectively inaccessible adjacent hexagons.

(Now, if these observations were deep mathematical insights worthy of publication in an eminent journal—or even a second-rate journal—we'd probably label the "Stark and Depressing Conclusion" *weak inaccessibility* and "Our Conjecture of Extreme Disconsolation" *strong inaccessibility*. Also, an aspect of the statement of the stronger result is worthy of comment: notice that the amount of detail accrued in the service of excluding unwanted interpretations renders it difficult to read and to understand. This seems to yield the counterintuitive notion that the more precise the transmission of an idea, the more opaque the language.)

Let us now rejoin our librarians, at the moment after their "stark and depressing conclusion."

Librarians Redux

It was as if our minds, mocking our exhausted, rooted, dispirited bodies, were set free. Almost in opposition to our wills, without fully digesting the realization, we continued to ruminate on these matters. One of us—does it matter which?—invoked a fragment in a contentious book found on a lower level, that read in part,

> Imagine a narrow flexible tube, one thousand miles long, called a "garden hose," laid out flat on a gigantic floor so that the hose impeccably fills the floor. The hose may curve abruptly, swirl painfully, spiral exuberantly, loop discursively, or even run straight, but it may never cross over itself nor rise from the ground in any way. Perhaps at many points the hose makes a kind of moral cusp or treacherous eddy and the close-by exterior parts of the tube nestle next to nearby parts of the tube. Skywards down to the hose, the view of the godlike will pinpoint many spots where the hose appears as parallel strands lying next to each other. At those spots, an ant—a tiny six-legged librarian—crawling through the interior of the hose may travel a considerable distance, perhaps miles, to reach a contiguous section. Even worse, never mind the Origin of the ant: the more it crawls, the more places it finds where the walls of the hose keep it further and further away from places the godlike can see. Lament, therefore, the linear forwards-and-backwards motion of the ant inside, while the nonlinear arabesques of the exterior

hose bring grace and redemption to those who can read them. Never shall the ant crawl from the interior and gaze upon the wholeness of the hose.

Paralyzed, we saw that although our limbs numbered four, and despite the fact that we weren't trapped in such a strange loop, there were striking similarities between the situation for the ants and for us. Regardless of the clever patterns taken by a godlike being laying down the garden hose, there must ever be more spots where the long, slimber structure of the loops of the hose would thwart an ant's attempt to move to any point athwart of the hose besides those immediately forwards or directly backwards. Clenching the hose into a crimp and then twisting it around in a whirlpool will produce a section where the ant could easily travel to all spots near its starting point, but then as the hose continues to be laid down, filling out the floor, circling around and again in a dizzying whorl of a world for the ant… we simply stopped talking, exhausted, looking up and down the airshafts.

Our Conjecture of Extreme Disconsolation: *There are unimaginably vast numbers of pairs of adjacent hexagons such that the span of our combined lives would not suffice to travel from the one to the other.*

Our earlier impotence was now seen to be a dream; our true plight lay revealed: perhaps we inhabited a section of the Library where all or most hexagons would allow us to attain only two of the six adjacent hexagons. All of those books, perhaps my or my friend's Vindication, perhaps a grammar of an ideal logic capable of straightening out the labyrinth in which we found ourselves, perhaps a fitting valediction for a carelessly dropped book mournfully hurtling down an airshaft, all these books would never be read by us.

SIX

More Combinatorics
Disorderings into Order

> *There is a secret element of regularity in the object which corresponds to a secret element of regularity in the subject.*
> —Johann Wolfgang von Goethe, *Maxims and Reflections*

> *Thinking man has a strange trait: when faced with an unsolved problem he likes to concoct a fantastic mental image, one he can never escape, even when the problem is solved and the truth revealed.*
> —Johann Wolfgang von Goethe, *Maxims and Reflections*

> *Either a universe that is all order, or else a farrago thrown together at random yet somehow forming a universe. But can there be some measure of order subsisting in yourself, and at the same time disorder in the greater whole?*
> —Marcus Aurelius, *Meditations*

CALCULATING THE NUMBER OF DISTINCT BOOKS in the Library, as seen in "Combinatorics: Contemplating Variations of the 23 Letters," is an example of a straightforward problem with a tidy solution. In this chapter, we do not so much solve a problem as explore how a maximally disordered and chaotic distribution of books in the Library can be seen as a Grand Pattern. This work is grounded in ancient ideas of combinatorial analysis, and although the ideas are consistent with the structure of the story, the ordering of the books we outline is incompatible with the Librarian's "elegant hope" that

> If an eternal traveler should journey in any direction, he would find after untold centuries that the same volumes are repeated in the same disorder—which, repeated, becomes order: the Order.

The Order conjectured by the librarian is an iterative order; a two-dimensional analogue may help to visualize it. Think of the complete ordering of all the books as being given by the imprint of a rubber stamp. After making an initial stamp (the section of the Library that the librarian lives in), without rotating the stamp at all and without overlapping stamps, continue applying the stamp up and down, left and right, and eventually cover the piece of paper. This translates the original order in all directions, vertically and horizontally, forming a simple kind of symmetry.

The Grand Pattern we propose in lieu of the Librarian's iterative Order is, in some sense, an ever-growing chain of concatenations of *all possible orderings*. To help envision what we mean, imagine that the Library is finite and approximately in the shape of a cube. Suppose we adjoined another Library-sized and Library-shaped building to the first one and distributed the $25^{1,312,000}$ unique books in a different ordering. This surely violates the Librarian's elegant hope, for it contradicts his vision that the addition should contain the books in precisely the same order as the original section. Now, suppose we continue to extend the Library by adjoining Library-sized and Library-shaped structures, each time distributing the books in a new ordering. Our endeavor now is to formalize the process, being as disorderly as possible, and at the end of a piece-by-piece construction, find an infinitely sized Library with a Grand Pattern occupying the whole of Euclidean 3-space.

Let's begin with a relatively simple question: how many distinct linear orderings are there of the three objects {▲, ■, ●} such that each object appears exactly once? A few moments of work produces the following list:

1. ▲, ■, ●
2. ▲, ●, ■
3. ■, ▲, ●
4. ■, ●, ▲
5. ●, ▲, ■
6. ●, ■, ▲

How might we convince ourselves that the list exhausts all possibilities? Perhaps by noting that we can fill the first slot three different ways, with either ▲, ■, or ●.

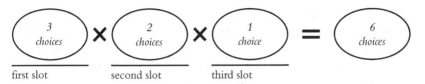

first slot   second slot   third slot

FIGURE 61. There are six ways to order three objects.

Once the first slot is filled, we are left with exactly two objects and two slots. Either of the two remaining objects can fill the second slot: we have two choices. Then, whichever object is left must fill the last slot. In other words, there are six different ways to fill the slots (figure 61).

Since the list has six distinct entries, we may be sure we've exhausted all possibilities. Generalizing this line of thinking, if we have four objects, there will be

$$4 \times 3 \times 2 \times 1 = 24$$

distinct ways to arrange the four objects: four choices for the first slot, three for the second slot, two for the third slot, and only one object remaining to fill the last slot.

Explicitly writing out the multiplications is viable for a relatively small number of objects. However, if we wished to signify the integer corresponding to the number of different ways to order only 25 objects, we'd find it cumbersome. Fortunately, a snappy notation, that of the *factorial*, was developed in the early 1800s:

$1! = 1$
$2! = 2 \times 1 = 2$
$3! = 3 \times 2 \times 1 = 6$
$4! = 4 \times 3 \times 2 \times 1 = 24$
$\vdots$
$25! = 25 \times 24 \times 23 \times \cdots \times 3 \times 2 \times 1$
$\vdots$
$n! = n \times (n-1) \times (n-2) \times \cdots \times 3 \times 2 \times 1.$

The orthographic symbol "3!" is read and pronounced as "three factorial," where "factorial" is understood to represent the process of multiplying an initial integer by every positive integer smaller than itself.

Several observations about factorials. First, on a personal note, even after 25 years of the serious study of mathematics we still tend to read "3!" as "THREE!" (very excitedly). Second, although we won't use the answers in this work, some natural questions to ask are "How is 0! defined?" and "Can we make sense of the expressions

$$\left(\frac{11}{2}\right)! \quad \text{or} \quad \pi!$$

even though they are not integers?" One tempting possible answer for the latter question is to focus on the "keep multiplying by numbers reduced in size by subtracting one" aspect of the factorial, and define, for example,

$$\left(\frac{11}{2}\right)! = \left(\frac{11}{2}\right) \cdot \left(\frac{9}{2}\right) \cdot \left(\frac{7}{2}\right) \cdot \left(\frac{5}{2}\right) \cdot \left(\frac{3}{2}\right) \cdot \left(\frac{1}{2}\right) = \frac{10,395}{64}.$$

Instead, in 1729, a similar yet more encompassing route was discovered by Leonhard Euler. Euler used the integral calculus to define a new function, called the *gamma function*, which, like the logarithm, possesses many interesting properties. One is that if a positive integer n is input to the gamma function, then $(n - 1)!$ is the output, meaning that the gamma function is essentially a generalization of the factorial. We test our naïve guess by inputting 13/2 to the gamma function, and find that the output is, in fact, pretty close:

$$\frac{10,395}{64}\sqrt{\pi}.$$

At any rate, factorial notation shares a property with exponential notation: it is easy to write down unimaginably large numbers. Take, for example, the number 70!, which by virtue of the simplicity of its written form appears as though it should fall within the grasp of the human imagination. In reality, 70! is larger than 10^{100}, and as we saw in the chapter "Combinatorics," 10^{100} grains of sand would completely fill 10 billion universes the size of our own.

Now, in the sort of action standard for a mathematician that incurs withering scorn from engineers, we accomplish the impossible by simply asserting it as a fact: number all the book-sized slots in the bookcases in the Library from 1 to $25^{1,312,000}$. (Wasn't that easy?) Via this numbering of the spaces in the bookshelves, we may use the factorial to compute the number of different ways to order the books in the Library. Even though the numbering of the slots in bookcases in the Library necessarily twists and snakes through three dimensions, we can still regard it as a consecutive sequence of $25^{1,312,000}$ slots laid out in a row. Put another way, regardless of how they are distributed in space, the positive integers have an intrinsic linear ordering, given by the progression 1, 2, 3, etc.

Given that each specific book fills a particular numbered slot, armed with the factorial notation we may trivially write down the number of different ways that the books in the Library may be shelved. Considering each book as a distinct object, there are

$$\left(25^{1,312,000}\right)!$$

different orderings. We'd like to get a sense of the magnitude of this number; after all, a factorial as small as 70! taxes our power of visualization by easily exceeding the number of subatomic particles in our universe. Fortunately, *Stirling's approximation to the factorial* gives a good estimation, in the sense that we can see this gargantuan number as an exponential of 10.

Stirling's approximation applied to $\left(25^{1,312,00}\right)!$ yields $10^{10^{1.834,103}}$.

This says that the number of different orderings of the books in the Library is approximately a 33-million-digit number; in the context of the story, it would take about 26 volumes simply to write down the number. The upshot is that Builders may construct a finite-sized Library housing all possible orderings of the books by assembling $\left(25^{1,312,000}\right)!$ Library-sized and -shaped buildings, filling each such building with exactly one ordering of the books.

If, though, along with the librarian, we assume that the Library is infinite in all directions, we have ample space for a more ambitious scheme than simply accounting for all orderings. We'll begin by defining a *libit* as a contiguous collection of accessible hexagons holding one particular ordering of the $25^{1,312,000}$ distinct books of the Library—any shape of the libit is acceptable so long as Euclidean 3-space can be completely *tiled* by replicas of that shape. Although we are imagining a libit looking roughly cubic or almost like a hexagonal prism, here's an extreme example unlike those: a tower of stacked single hexagons sufficient to hold all the distinct books. Here's another: a giant near-hexagon of hexagons completely contained on one floor, again sufficiently large to hold all the distinct books.

Next, tile all of 3-space by clones of the libit, and from the infinite possibilities, arbitrarily choose an initial libit and also any hexagon contained within. We'll use this hexagon as a reference point, and consider it to be the origin of the Library. Starting at the origin, successively choose contiguous hexagons in an orderly fashion until the first libit is completely numbered. (It is not unreasonable to worry about how to ensure that every hexagon will be numbered. We address this issue in the chapter "Critical Points.")

Now choose an adjacent libit, and starting in the "same" hexagon as in the first libit—that is, in the hexagon in the second libit that corresponds to the origin—extend the numbering starting with $25^{1,312,000} + 1$. The numbering of the second libit will, of course, run all the way up to $2 \cdot \left(25^{1,312,000}\right)$. Next, repeat the process by choosing a third libit contiguous to the second one, and number the slots on the shelves as before.

Continue to iterate the procedure until the shelves in $\left(25^{1,312,000}\right)!$ contiguous libits are numbered, and note that each of the $\left(25^{1,312,000}\right)!$ different orderings is composed of $25^{1,312,000}$ different books. Thus, for the first step of the Grand Pattern, we utilize the first

$$25^{1,312,000} \times \left(25^{1,312,000}\right)!$$

= (Number of distinct books) × (Number of distinct orderings)

slots in the infinite Library.

However, as we were filling the libits with different orderings, we were implicitly making choices regarding the possible orderings of the books. Making the Grand Pattern requires us to leap categories and consider orderings of orders. Let's do a smaller-scale example of a Grand Pattern constructed in the Euclidean plane.

Instead of books, we'll use the three letters {a, b, c}. In Step 1, we give a straightforward ordering of the three letters, which, by analogy, is similar to using books to fill the first libit. For Step 2, we produce one list of the 3! = 6 possible orderings of the three letters, and we think of this as the set of $\left(25^{1,312,000}\right)!$ libits described above (figure 62). Next, note that there are 6! = 720 distinct lists of six orderings. This is exactly the

MORE COMBINATORICS ~ 113

Step 2		
a	b	c
a	c	b
b	a	c
b	c	a
c	a	b
c	b	a

Step 1

a	b	c

Here are Steps 1 and 2 of a Grand Pattern constructed in the plane from the three letters a, b, and c.

Step 1: This is one possible ordering of the three letters.

Step 2: This is one possible ordering of the list of all six different orderings of a, b, and c. Note the ordering from Step 1, which is subsumed in Step 2, is shown in gray.

FIGURE 62. The beginning of a Grand Pattern.

idea we want to communicate: that although we ran through all possible orderings in the list of six, if we think of each of the six orderings as a new unit there are 720 orderings which must be accounted for in Step 3. Put another way, first we worried about all the ways to order {a, b, c}. Now, we want all possible orderings of

{ [a, b, c]; [a, c, b]; [b, a, c]; [b, c, a]; [c, a, b]; [c, b, a] } .

Figure 63 shows five distinct lists, beginning a spiral, which ultimately creates a new, larger rectangle filled out by all 720 distinct 3 × 6 rectangles. This, in turn, is an ordering of orderings of orderings and guarantees that at the next stage, Step 4, we'll be able to continue to spiral around and create an even larger rectangle.

Now, let's move out of the plane, return to the Library, and apply these ideas to create the Grand Pattern there. At the conclusion of the first step, we went on hiatus having filled $(25^{1,312,000})!$ libits, each representing an ordering of the Library. Now we boldly expand the Library until there are

$$((25^{1,312,000})!)!$$

libits, each one representing a distinct ordering of distinct orderings. In other words, this second step of the Grand Pattern subsumes the first step as just *one* particular ordering of all orderings of the books of the Library.

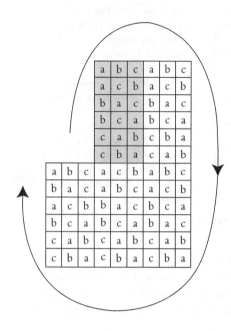

Step 3: Here's the beginning of one possible ordering of the 720 orderings of the lists of six distinct orderings of the letters a, b, and c. (If this picture were complete, there would be 3·6·720 = 12,960 boxes with letters contained inside: all possible ways to write lists of six triplets of letters.)

Pictured are five distinct lists, each of which is a legitimate example of a list satisfying the needs of Step 2. Note that the additional lists form a spiral around the first, and it is essential that the collection of all 720 lists ultimately forms a rectangle.

Continuing in this fashion through subsequent steps eventually fills the plane with all orderings, orderings of orderings, orderings of orderings of orderings, etc. (Step 4 would consist of 3·6·720·720! boxes with letters contained inside. This is approximately $10^{1,751}$ boxes.)

FIGURE 63. Taking another step towards the Grand Pattern.

And now the iterative process grows clear; the third step of the Grand Pattern must account for all orderings of all orderings of all orderings of the books in the Library. There are

$$(((25^{1,312,000})!)!)!$$

such orderings. The third step of the Grand Pattern subsumes the second step as just *one* particular ordering of all orderings of all orderings of the books of the Library. And so on, and so on, and so on.

One appealing aspect of the Grand Pattern is that for any conceivable finite assemblage of orderings in libits, infinitely many sections of the Library will contain precisely that same distribution of books. Each new step incorporates the preceding step as one subunit of the new step, and so each step repeats and repeats and repeats.

Another consequence, somewhat amusing, is that there will be abutting libits whose distributions are, in a sense, palindromic. A librarian, fortuitously born by the border between these two libits, would find a vast "wall" of adjacent duplicate hexagons! Moving in one hexagon away from the border in each direction would also reveal duplicate hexagons. Ditto for moving in two hexagons away from the border, and so on. There are as many quirky distributions in bordering orderings as we can imagine; after all, every distribution appears next to every other. Could a human librarian born in such a border locale guess that the Library contained volumes consisting of all possible variations of 25 letters?

A sharp-eyed reader may have noticed that we've consistently written "the Grand Pattern," as opposed to "a Grand Pattern." This is because once we have settled on a libit shape, *there is only one such Grand Pattern*. (See the Math Aftermath for more about why we must choose a libit shape.) Here is one way of seeing this mild form of uniqueness of the Grand Pattern:

Assume we've already constructed an infinite Library with books distributed in a Grand Pattern. Also assume, for the sake of argument, that a godlike entity also constructs an infinite Library, and for reasons that range from the puckish to the profound, wishes to distribute books in a *different* Grand Pattern. This Other Entity chooses the same-shaped libit, and after envisioning a tiling of the second Library, starts at an arbitrary hexagon, and for the first step, distributes $25^{1,312,000}$ books in an allocation varying from ours. Next, for Step 2, the Other Entity distributes books into the remaining

$$\left(25^{1,312,000}\right)! - 1$$

libits in a distribution unlike our second step. And so on.

Let us add omniscience to our list of godlike attributes. Consequently, we *know* exactly how the Other Entity will distribute the books for the first step, the second step, etc. etc. Since every finite pattern, no matter how large, appears in our Grand Pattern, we simply choose an initial libit from our pattern which has the same distribution of books as the Other Entity's first step. By dint of omniscience, in fact, we chose that initial libit so that, moving outwards, it exactly shadows the Other Entity's second step, too. In fact, for any positive integer n, we chose so well, that moving outwards shows precisely the same distribution as the Other Entity's nth step; and remember, as implausible as this may seem, we must remind ourselves that all finite patterns, no matter how large, appear in a Grand Pattern. Finally, we observe that since our Grand Pattern exactly mimics the Other Entity's Grand Pattern for every positive integer n, it shadows it for all n; therefore, the two Grand Patterns must be the same. (See the Math Aftermath for more about this leap of logic from the finite to the infinite.)

A librarian granted eternal life and a Funes-like infinite capacity for photographic memories, walking a Grand Tour of the Library, reading and distinguishing every book, would discover that no pattern reliably repeats. On the other hand, a librarian of genius, equally endowed with the unimaginable ability to process titanic amounts of information, might well guess that, for a choice of the shape of a libit, the books are distributed throughout the Library in all possible orderings. And all possible orderings of all possible orderings. And all possible orderings of all possible orderings of all possible orderings...

As such, the books of this infinite Library are maximally disordered; and yet this ultimate disorder forms a unique overarching Order of all orderings: the Grand Pattern.

Math Aftermath: Libits, Uniqueness, and Jumping from the Finite to the Infinite

> *It is hard to be finite upon an infinite subject, and all subjects are infinite.*
> —Herman Melville, *The Piazza Tales and Other Prose Pieces*

In what follows, we briefly discuss two subtle points from above. First, we examine why the libits must be the same shape while comparing their respective Grand Patterns. If they are allowed to be dissimilar, we'll use, as an example, the two extreme examples from earlier in the chapter: we'll say a *tower libit* is a slender tower of stacked single hexagons, while a *floor libit* is completely contained in one floor. Imagine seven tower libits, six of them adjacent to a central one, that are exactly the same save that in their top hexagons, four books are permuted into slightly different positions. Then the towers are all identical except in the top hexagons, and even there, each top hexagon contains the same books as the others. (See figure 64.)

We claim that no floor libit can accommodate this distribution, for any floor libit that contains any hexagon of the central tower libit necessarily also contains a hexagon of at least one of the adjacent tower libits. By construction, these two hexagons must contain precisely the same books. However, no libit can have duplicate books; thus it follows that a Grand Pattern constructed out of tower libits cannot be replicated by floor libits. (See figure 65.)

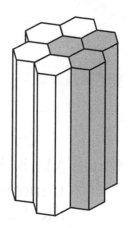

FIGURE 64. The seven adjacent tower libits.

In this part of the Math Aftermath, we consider the leap from the finite agreement at "each n" to the infinite agreement at "every n" that arose in the comparison of our Grand Pattern and the Other Entity's Grand Pattern. Here is a sequence of ideas that may help act as a bridge across the unimaginable abyss between the finite and the infinite: suppose our Grand Pattern differed from the Other Entity's. Then, by the well-ordering principle (see below), there is a smallest positive integer—for example, 412—such that our Grand Pattern differs from the Other Entity's Grand Pattern at Step 412. However, we know that the Grand Patterns are exactly the same at each finite step, including step 412. Thus, there can be

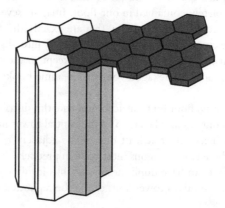

FIGURE 65. Any floor libit intersecting the central tower.

no smallest integer at which they differ. Therefore, they are everywhere the same!

The *well-ordering principle* is an axiom of set theory (in some less commonly used logical models of arithmetic, it is a provable theorem). It says, in essence, that any set of positive integers has a smallest member. Although this sounds relatively innocuous, the constructivist school of mathematicians raises nontrivial objections to the well-ordering principle and the theorems which spring from it. It is worth mentioning that although these objections can never be formally refuted, there are few working pure mathematicians who are constructivists. On the other hand, a number of prominent computer scientists and applied mathematicians are in sympathy with the philosophic beliefs of the constructivists; for example, for them, a number doesn't exist unless it can be constructed. A strict constructivist won't admit the actual existence of π or $\sqrt{2}$ in the sense of an endless decimal expansion; rather, such a thinker would only acknowledge rational approximations to these irrational numbers.

SEVEN

A Homomorphism
Structure into Meaning

Because we do not understand the brain very well we are constantly tempted to use the latest technology as a model for trying to understand it. In my childhood we were always assured that the brain was a telephone switchboard. ('What else could it be?') I was amused to see that Sherrington, the great British neuroscientist, thought that the brain worked like a telegraph system. Freud often compared the brain to hydraulic and electromagnetic systems. Leibniz compared it to a mill, and I am told some of the ancient Greeks thought the brain functions like a catapult. At present, obviously, the metaphor is the digital computer.

John Searle, *Minds, Brains and Science*

What is a divine mind? the reader will perhaps inquire. There is not a theologian who does not define it; I prefer an example. The steps a man takes from the day of his birth until that of his death trace in time an inconceivable figure. The Divine Mind intuitively grasps that form immediately, as men do a triangle. This figure (perhaps) has its given function in the economy of the universe.

—Jorge Luis Borges, "The Mirror of Enigmas," in *Labyrinths*

THE TIME HAS COME TO RETURN TO THE FIRST person singular voice. I've done my best, up to this chapter, to restrict myself to excavating, extending, enlarging, and explaining mathematical ideas which naturally arise from the story. But now, after scanning critical reviews and interpretations of coded meanings of the story, I find I have another reading to add to the existing constellation: a homomorphism between the structures of two different ideas. In biology, a homomorphism is a correspondence in appearance or form, but not in structure or origin. In mathematics, a homomorphism has almost the exact opposite meaning: it is a formal *map* between two seemingly dissimilar algebraic objects that illuminates a deep correspondence between their underlying structures. Unfortunately, interesting examples require background information beyond the scope of this book, and I stress that I am using the term "homomorphism" metaphorically. (In fact, I'm using "homomorphism" as a synonym for "metaphor.")

In 1931, by proving his first and second incompleteness theorems, Kurt Gödel inspired a thorough—and ongoing—reconsideration of the logical foundations of mathematics. Many excellent books have been written on Gödel's theorems; for my purposes, it is enough to think of the first incompleteness theorem as saying that if a formal axiomatic system is rich enough to generate the positive integers and the operations of addition and multiplication, then there are statements expressible that are *undecidable*: they cannot be proven true within the system of axioms, nor can they be proven false. In other words, the system is *incomplete*. (The second incompleteness theorem is a deep related result about proving the consistency of such axiomatic systems.)

In 1936, Alan Turing became a founder of computer science when he creatively combined his own brilliant ideas with some of Gödel's and published "On computable numbers, with an application to the Entscheidungsproblem." The *Entscheidungsproblem* was originally posed by Leibniz in the 1600s and formalized by Hilbert in 1928. It wondered whether a mechanical process or machine, manipulating symbols, could correctly assign truth values to statements posed within an axiomatic system such as mathematics. In other words, the problem asked if it is possible to create an automatic process that would determine whether any given theorem is true or false. The mechanical process or machine wouldn't need to provide a proof of a true theorem; it simply needed to correctly identify it as true.

Turing's brilliant tactic was to focus on the halting problem for computers, which hinges on a simple yes/no question: given a program coupled with a starting condition for that program, once it has begun to run, does it ever stop? For example, a program might be called NextPrime, and depending on the starting condition, the output will be the very next prime. If we use 17 as the input, then NextPrime will output 19, as 19 is the next prime after 17. On the other hand, using the current state of number theory and computer technology, if $25^{1,312,000}$ is input to NextPrime, the program would run for years without giving the next largest prime. In this context, the Halting Problem asks if there could be one master program called, say, HALT, which would take the program NextPrime and starting condition $25^{1,312,000}$ as inputs, and would output YES—because, in fact, given sufficient time and resources, NextPrime will eventually output the first prime larger than $25^{1,312,000}$.

Let's consider the alternative, a program that never halts. An example of such a program involves finding pairs of *twin primes*, which are two prime numbers p and $(p + 2)$. For example, $\{3, 5\}$ and $\{29, 31\}$ and $\{59, 61\}$ are pairs of twin primes. The Twin Prime conjecture states that there are infinitely many pairs of twin primes, and in the past 100 years, mathematicians have amassed a fair amount of evidence to suggest that it is true. On the other hand, no proof has yet been found, so it is possible that there is an unimaginably large pair of twin primes that is the last such pair. A point of interest is that the Twin Primes conjecture is not falsifiable by showing the existence of a huge stretch of consecutive primes that have no twins. It's always possible that that the very next prime will have a twin.

Let TwinPrime be a program that finds twin primes, and let the starting input be the prime number two. When TwinPrime starts, it is guaranteed to run forever, either outputting pairs of twin primes until the computer crumbles into dust or, if there is a largest pair of twin primes, continuing to fruitlessly check larger and larger primes to see if they have twins. Whether or not there are a finite number of twin primes is irrelevant: TwinPrime will never halt. Thus, if HALT existed, if we input the program TwinPrime and a starting condition into HALT, the output would be *NO*—this program will never halt.

However, Turing showed that the halting problem is undecidable; that is, there is no way to construct a master program HALT that will always be able to determine if an input program and starting condition will or will not halt. As a consequence of the undecidability of the halting problem, it follows that the answer to the Entscheidungsproblem is negative: there can be no such process or machine that will decide whether a theorem in arithmetic is true or false.

A key element of Turing's proof is the creation of an abstract computing machine, since dubbed a *Turing machine*. The Turing machine consists of a black box and an infinitely long strip of tape; if the concept of "infinitely long tape" is distressing, instead think of it as "a very long tape, with dedicated crews of tape-extenders who are always able to add

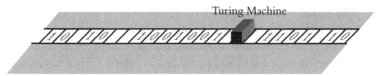

FIGURE 66. A Turing machine sits on an infinitely long strip of tape.

more tape as necessary." The tape is divided into little one-unit squares, and the black box occupies one square at a time. Each square of the tape may contain an orthographic symbol from a finite alphabet on it, or it may be blank (figure 66).

The black box begins sitting at a given spot on the tape, called the *initial position*, and the black box is in an initial *internal state*. An internal state is a simple instruction on what to output, given a particular input received. There can only be a finite number—possibly very large—of internal states. The black box reads the symbol in the square that it's sitting on, and this is the input for the Turing machine.

The Turing machine may then do several different things, depending on its internal state and the symbol in the square:

- It can erase the symbol in the square.
- It can write a new symbol from the finite alphabet in the square.
- It can change to a new internal state.
- It can move either to the left or right one square.
- It can halt (permanently).

That's it! Despite their prosaic description, Turing machines are remarkable. Although the Turing machine would run much, much, much slower, a programmer could write instructions for it that would reproduce any program available on the most advanced supercomputer in existence. Every single program for a desktop computer, including any program for word processing, graphics, or internet browsing: all may be converted into a format suitable for a Turing machine, which would then produce exactly the same output. (Of course, without a suitable way of converting the infinite tape's information into a picture on a screen, the output would be difficult to recognize.) Stronger still: any computing task currently imaginable by human intelligence may be performed by a suitably programmed Turing machine.

The Church-Turing thesis captures this state of affairs and projects it into the future; it says that *any* task done by *any* possible computing device running *any* possible software could also be done by a suitably programmed Turing machine. The reason it is called a "thesis" as opposed to a "theorem" is that it can never be proved. There's simply no way to do so, because we have no idea of the possible computing devices or programs that might be dreamt of and implemented at some time in the future. On the other hand, the Church-Turing thesis could be disproved, because someone may invent a new kind of computing device which performs a task beyond the capabilities of a Turing machine. For a few decades, logicians sought to disprove the Church-Turing thesis, but every attempt failed. Thus, the default intellectual position is that the Church-Turing thesis is plausible, and it continues to gain in plausibility as the years go by; the longer it hasn't been disproved, the more plausible it becomes. A veritably Borgesian state of affairs: we know that we don't know and even know precisely that we might never know.

This, then, is the milieu of the Turing machine, the dizzying maze of ideas it inhabits: a welter of halts, non-halts, infinite loops and regresses, computations, equivalences, incompletions, logical indecisions, definitive answers, and potentially eternally unknown truths. These ideas formed part of the Zeitgeist of the 1930s, and while they would not have appeared in any popularization of mathematics, it is probable they entered the general intellectual discourse of the era, and therefore possible that echoes reached Borges.[1] Regardless of Borges' conscious knowledge, I propose the following homomorphic mapping: librarians to Turing machines.

A librarian is constrained to move from one room to another. As a careful reading of the description of the Library indicates, a passage through the hexagons on any given floor constitutes a path which allows only the possibility of forward or retrograde motion. (It doesn't matter that the librarian can take stairs up or down; it's been demonstrated that Turing machines with multiple tracks and switchings are equivalent to the one-track version.) Each room is filled with a particular collection of symbols from a finite alphabet, which the librarian reads. The librarian, depending on the librarian's internal state and the books in the hexagon, may then either

- Erase the symbol in the square; that is, for example, throw the books down the airshaft.
- Write a new symbol—from the finite alphabet—in the square; that is, either reorder the books in the hexagon, move books from hexagon to hexagon, or actually, as the narrator of the story states, inscribe symbols on the flyleaf or in the margins of a book.
- Change to a new internal state; that is, the librarian can go to the bathroom or go to sleep. Or, perhaps the librarian has a revelation and by acquiring new ideas, moves to a new cognitive view of the world. Or even simply that the librarian's mood changes.
- Move to the left one square or move to the right one square; that is, move to a new hexagon.
- Halt permanently; that is, expire.

The librarian's life and the Library together embody a Turing machine, running an unimaginable program whose output can only be interpreted by a godlike external observer.

A user.

A reader.

EIGHT

Critical Points

In the mountains of truth you will never climb in vain: either you will already get further up today or you will exercise your strength so that you can climb higher tomorrow.
—Friedrich Nietzsche, *Maxims,* aphorism 358

CRITICAL POINT IS A TERM THAT HAS ASSUMED A host of meanings in mathematics. Generally, it denotes a location where typical behavior breaks down and unusual and interesting phenomena occur. In the qualitative study of the solutions of differential equations, some critical points are singularities where flow lines of solutions may converge from many different regions and then, regardless of initial proximity, shoot off in a variety of divergent directions. This seems an apt metaphor for critical perspectives arising from the interpretation of a literary work.

Taking advantage of Nietzsche's metaphor, at the base, mountains offer a multiplicity of approaches towards the summit. Frequently these trails converge to several natural routes, and oftentimes the descent must be taken via an unexpectedly different path. Thus it is no surprise that in my quest to read all commentaries by critics on "The Library of Babel," I learned that many predecessors have independently climbed and descended the mountain, some along paths with sections that closely parallel mine. This chapter is devoted towards outlining some of these trails. In particular, I am restricting myself to those that involve mathematics in one form or another. I begin by acknowledging those who independently found some of the same mathematics in the story.

The eminent mathematician and pioneer German science fiction writer, Kurd Lasswitz, in his 1901 story "The Universal Library," not only calculates the number of books in his universal library, but also mentions that filling our known universe with books barely dents the total. As mentioned in the first endnote to chapter 1, Amaral, Rucker, Nicolas, Faucher, and Salpeter all calculate the number of books in the Library, while Bell-Villada excerpts a passage from Gamow's *One, Two, Three... Infinity* that shows he has a notion of how such a calculation should be performed.

While looking for a reference, for the interested reader, to the topology and cosmology of the Library presented above, I discovered that in their 1999 article "Is Space Finite?," Luminet, Starkman, and Weeks listed "The Library of Babel" as a suggestion for further reading. At least one of them must have thought about the topology of the Library. Floyd Merrell also speculates about the topology of the Library and briefly discusses the possibility of a catalog, albeit mostly in the context of space-time physics, when he talks about light-cones for librarians and their world-lines, which are infinitesimal by contrast to the size of the Library. Furthermore, inspired by work of Bernadete, Merrell includes a brief discussion of a Book of Sand that is similar in spirit to my first interpretation.

Whereas I have mainly sought to elucidate the mathematics in the story, most commentators endeavor to use mathematics to create a framework of analysis of the story and, more generally, Borges' oeuvre. It strikes me as an interesting philosophic pursuit to examine the project of importing mathematical and scientific terminology and systems into the field of literary analysis. I recuse myself from this study on the grounds that I am professionally neither a philosopher nor a critic. I am, however, qualified to comment on the correctness of the mathematics brought to bear on Borges, and, by dint of careful and extensive reading of Borges, to agree or disagree with various other interpretations of "The Library of Babel."[1]

Given that literary critics mount defensible arguments for the primacy of interpretation over authorial intent, why should we worry whether or not a critic's use of mathematics is perfectly correct? By way of an answer, consider the following hypothetical misreading of Borges. The not-so-eminent critic William Goldbloom Blockhead, my not-so-bright alter ego, has confused "South America" with "South Africa." After all, they sound similar and are both, more or less, continents. Blockhead's stirring postcolonial analysis of Borges' work raises new and disturbing questions. For example, Blockhead wonders why Borges wrote in Spanish in lieu of English or Afrikaans, declines to mention apartheid, and exercises narrative space on Argentine border and independence wars rather than the Boer War and World War I. Does this mean that Borges' own political views must be coded, rather than overtly expressed? Next, Blockhead points out that the virtual invisibility of Africans in Borges' writing is a strong sign that, in fact, race is his most important agenda—what else could explain such an egregious omission? Blockhead's analysis triumphantly concludes with the unique insight that this interesting *homme de lettres* employs these strategies to avoid the literary trap of being read as merely a colonist writing in and about a colony.

It's conceivable that some of Blockhead's remarks may be of independent interest, and some may even apply to Borges—after all, there are some real political and historic parallels between Argentina/Spain and South Africa/England. Nevertheless, one may well feel an irresistible urge to ignore wholesale the stream of mistaken inferences following from Blockhead's wrong assumption. Seen in this light, then, I offer bouquets of refractions arising from the light of other critic's works.

Sarlo, in *Jorge Luis Borges: A Writer on the Edge*, tellingly sees the Library as a symbol of political oppression in the milieu of a totalitarian state. The librarians are, in today's vernacular, information serfs who will never be able to acquire the necessary data to transform their status. She writes that

> Structurally, the Library is also a panoptic, whose spatial distribution of masses and corridors allows one to see every place in it from any of its hexagons. The panoptic design of the Library brings to mind that of a prison where the guards should be able to see any cell from every possible perspective. Foucault has studied this layout as a spatialization of authoritarianism, as an

image of a society where total control is possible and no private place (no private thought) is admitted. The universe described as the Library lacks any notion or possibility of privacy: all the activities are, by definition, public.

The mental exercise implicitly asked now of the Reader is to imagine oneself in the Library and intuit the sensory and emotional experience: Is it dimly or brightly lit? Are the books musty or shockingly pristine? Would the airshafts induce a tremendous vertigo or could they be overlooked? Do the spiral staircases, placed either at every entrance or every other entrance, hem librarians as do the bars of a cage, or are they comforting mileposts along their life paths? Are there doors on the small rooms designed for sleeping and physical necessities, or have the librarians grown up acculturated to a different kind of privacy than us?

Once we establish our imagination in one of the hexagons, we see that the Library is not panoptic; the only available line of sight is in the air shaft central to the hexagon. Short of sticking a head into the airshaft while looking up or down, only a few hexagons would be visible before the convergence forced by the rules of perspective would hide all but a small part of the floors and ceilings.

Furthermore, as discussed in the chapter "Geometry and Graph Theory," it isn't clear what the sight lines are like on an individual floor; hexagons may be arranged in a straight line, or the entrances of the hexagons may well curve around nonlinearly. The most extensive vision afforded by the structure of the Library, then, would be a hexagon at the intersection of a cross formed by its airshaft and a straight-line corridor. However, even if a passage ran straight, the spiral staircases would block the views. By contrast, Bentham's Panopticon, as described in Foucault's famous work *Discipline and Punish*, enables full-time viewing of all inhabitants by obscured central scrutinizers: an altogether different geometry.

Sarlo also writes that "the infinity of the Library cannot be empirically experienced, even if a traveller were granted infinite time." Even if the Library extends forever in the manner of Euclidean 3-space, there is a nifty way to see that the whole Library may, in fact, be visited. First, we make the reasonable assumption that the architecture of the Library allows all adjacent hexagons to be visited. Now, imagine starting in any hexagon, which we may now consider to be the Origin of the Library.

1. Begin at the Origin and visit every hexagon adjacent to the Origin on the same floor as the Origin. Proceed up one floor and pass through all the hexagons on this floor that are above a previously visited hexagon. Now, go down two floors to the floor below that of the Origin and visit every hexagon on this floor that is below a previously visited hexagon. It is the case now that every hexagon a "distance" of one hexagon from the Origin has been visited.
2. Next, beginning on the floor of the Origin, visit every hexagon adjacent to a previously visited hexagon. Hence, on the floor of the Origin, all hexagons within two hexagons of the Origin have been visited. Now use the spiral staircases to proceed up *one* and *two* floors and visit all the hexagons on them that are above a previously visited hexagon. Next, go down one floor below the Origin, then two floors below the Origin, and do the same. Note that every hexagon a "distance" of two hexagons from the Origin has been visited.
3. Next, again starting on the floor of the Origin, visit every hexagon adjacent to a previously visited hexagon, then proceed up *one* and *two* and *three* floors, visiting all the hexagons on each of these floors that are above a previously visited hexagon. Do the same for the one, two, and three floors below the floor of the Origin, and it follows immediately that every hexagon a "distance" of three hexagons from the Origin has been visited.
4. Etc.

If this hexagon-visiting algorithm is carried out, it is not hard to see that at any stage, only a finite number of hexagons have been visited, and a traveler granted infinite time must eventually visit every single hexagon. This last assertion follows because even if the Library extends infinitely in all directions, it still must be the case that any hexagon in the Library is fixed at a finite number of hexagons from the Origin—the hexagon in which the traveler began.

Although he never explicitly mentions "The Library of Babel," I include a discussion of Svend Østergaard because I believe the book proposed in the final footnote of "The Library of Babel" is of a similar structure to the book described in Borges' short story "The Book of Sand." In *The Mathematics of Meaning*, Østergaard discusses the Book of Sand, but proceeds under the unwarranted assumption that it possesses uncountably infinitely many pages.[2]

I find it unlikely that there are uncountably many pages, for the narrator of "The Book of Sand" only mentions the finding of integer-numbered pages: if that was so, it would be extraordinary to find even a single page numbered by an integer, because the likelihood of randomly finding an integer in the real number line is *zero*—or if that sounds improbably absolute, "vanishingly small." This is because the set of integers is countable and therefore, when considered as a set contained inside of the real numbers, it is of measure 0. This entails that any integer is much harder to find than a single prespecified dust speck adrift in South America. (See my second interpretation of the Book of Sand in the chapter "Real Analysis" for a discussion on measure 0.)

This is essentially the reason that while the narrator for "The Book of Sand" is looking at a particular page, the mysterious stranger adjures him to "Look at it well. You will never see it again." The probability of randomly picking the same integer twice is also vanishingly small. To see this, imagine opening the Book of Sand to page 17. If there were only 100 pages in the Book, each time it was opened again there would be a 1/100 chance of randomly opening it to page 17. If there were 1,000 pages, there would be a 1/1,000 chance. If there were a million pages, there would be a 1/1,000,000 chance. If there were infinitely many pages, it is tempting to write that there would be a $1/\infty$ chance, meaning "probability 0." But it wouldn't be correct to write that, and the story of the probability, while interesting and exciting, is beyond the scope of this book. (If the stranger and narrator were truly interested in seeing a page a second time, it's fair to wonder why they didn't simply insert a cardboard bookmark to reenter this bookish Heraclitean river twice.)

Moreover, the narrator of "The Book of Sand" states that illustrations occur every 2,000 pages. If there were uncountably many pages—that is, the same number of pages as there are of points in the real number line—then there would be no way of counting the number of pages between two selected pages. In fact, there'd be no way to find a "next page," for in the real number line, numbers lap up against each other with no "closest" number. This is known, in various guises, as the *Archimedean property*.

I have great sympathy for the last two critics I'm going to discuss, N. Katherine Hayles and Merrell; their project, as I understand it, is truly noble. They seek to create or expand upon a theory which accounts for all the complex interrelations between the perceivable universe, consciousness, all previous and current human works, the Zeitgeist, culture, language, author, text, interpretation, and reader. Such a theory would, by virtue of absorption, dwarf a Grand Unified Theory of Everything from physics. It is natural, therefore, that two literary critics, steeped in the disciplines of chemistry and physics, would appropriate the language and approaches of mathematics and science to employ them in this most ambitious theory.

Hayles' work (Hayles, 138–67) primarily consists of associating ideas of self-referentiality and infinite sequences, infinite series, and infinite sets to Borges' work. Many of her insights are deep. Although some passages seek to persuade the reader of the meaninglessness and marginalization of mathematics, Hayles is content to use mathematics as a means for understanding Borges, perhaps in the same way a sponge, riddled with holes, is useful in sopping up fluid reality.

After a précis of the story, on page 151 Hayles critiques the librarian's "elegant hope" by noting that "the narrator's 'solution' is of course an answer only in a very narrow sense. While it suggests a way to transform randomness into ordered sequence, it contains no hint of how that sequence may be rendered intelligible or meaningful." As I noted in the chapter "Topology and Cosmology," the patterning of a periodically repeating Library may be thought of as symmetric three-dimensional wallpaper. For example, the illustration in figure 67 is, in some sense, random and chaotic. However, when it repeats periodically, it takes on a pleasant enough symmetry; an order, if you will (figure 68). I contend again that this is the Order that in-formed the narrator's out-look.

FIGURE 67. A random, chaotic sketch.

Hayles' main intent, in her reading of "The Library of Babel," is poetic: she wants a Borgesian "Strange Loop" to dissolve the boundary between the reader and the text by roping the reader into the story itself. However, to accomplish this lyrical agenda, Hayles writes on page 152 that "Logic demands that we conclude the present text in hand (which of course is printed) to be the Library's book. What we have is not the narrator's handwritten text but a mirror of it, or perhaps one of the 'several hundreds of thousands of imperfect facsimiles.'" It is curious that a critic eager to limit logic should invoke it almost as a magic amulet, for "logic" doesn't "demand" anything. Rather, it seems to me that Hayles is attempting to have her theory of Strange Loops produce a variation of a result that Borges himself stresses in the story, "This useless and wordy epistle already exists in one of the thirty volumes of the five shelves in one of the uncountable hexagons—and so does its refutation." The story is in the Library, the book it originally appeared in, *Ficciones*, is in the Library, and the complete works of Borges are in the Library. Hayles' books, the words of this book, and anything that can be written using

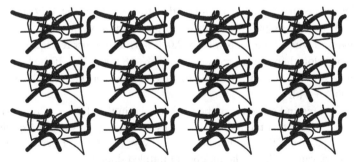

FIGURE 68. Wallpaper formed by periodically repeating a random, chaotic drawing.

25 orthographic symbols: all are necessarily in the Library. Nevertheless, none of these inclusions implies that we are at this moment reading a Library book or that we are librarians roaming a universe of hexagons.

If we can legitimately *assume* as a *premise* that we are holding a Library book in our hands, then Hayles' next set of ideas, which are intriguing, do follow: "... even more important is the implication that we are reading the Library's book. This, in turn, implies that we, like the narrator, are within the Library examining one of its volumes, which means that we, no less than the narrator, are contained within one of the books we peruse." Since the premise is unfounded—my copy of *Ficciones* is not 410 pages, I'm not in a dimly lit hexagon—the chain of implications does not follow.[3] If, on the other hand, Hayles was referring to a sort of narratological space created by the story, where we readers accept that by virtue of reading the story we are somehow in the story's confines, then it still doesn't follow that we are reading a printed text of the Library rather than a handwritten note of an avuncular librarian. In fact, given that we are human, inhabiting a miniscule section of the Library where humans reside, it is vastly more likely that we would stumble across a book which a human has inscribed than one which contains the story "The Library of Babel."

The story, including spaces, is comprised of approximately 18,000 lexical symbols. These contiguous 18,000 symbols could occupy 1,294,001 different starting positions in a 410-page book. To see this, observe that the first such position would entail that the first symbol of the story occupied the first slot of the book's 1,312,000 slots. The last such position has the concluding period of the story occupying the last slot in the book. This entails that the first symbol of the story occupies the 1,294,000th slot. It may be helpful to visualize this process as a block of 18,000 red squares moving along a tape of 1,312,000 slots. If the first red square is at the first slot, then the last red square is at the 18,000th slot. If the last red square is at the last slot, then the first red square is at the [(1,312,000 − 18,000) + 1]th slot.

If we remind ourselves of the work we did in the chapter "Combinatorics," we see that the number of books containing the story *at a specific starting position* is the number of different ways the slots not occupied by the story can be filled. The number of unfilled slots is

$$25^{1,312,000-18,000} = 25^{1,294,000}.$$

From the first paragraph of this note, we see there are 1,294,001 such positions; thus there are approximately

$$(1{,}294{,}001) \cdot (25^{1{,}294{,}000})$$

distinct books in the Library containing the story. So the probability of finding a Library volume containing the story is

$$\frac{\text{(books we care about)}}{\text{(total number of books)}} = \frac{(1{,}294{,}001) \cdot \left(25^{1{,}294{,}000}\right)}{25^{1{,}312{,}000}}$$

$$= \frac{1{,}294{,}001}{25^{18{,}000}},$$

which, in turn, is approximately

$$\frac{1}{10^{25{,}157}}.$$

This is roughly equivalent to the likelihood of winning a major lottery 3,600 times in succession!

There is, however, a profound sense in which Hayles is correct, a sense that Borges explicitly intended. You *are* in the Library. A multivolume set is scattered throughout the Library that details every single day of your life, including your death. In fact, a multi-quintillion volume set that details the lives, deaths, and protein transfers of each and every one of your cells is also scattered throughout the Library.

By invoking this theme in conjunction with the idea of potential inaccuracies of a particular volume, Hayles opens the door to a stimulating line of thought. Suppose, miraculously, you were to find a grouped set of volumes, each of which had one page dedicated to one day of your life. Every single page, as far as your memory can recall and corroborate, is an accurate portrayal of that day. You read the page that corresponds to *tomorrow*. At the end of the next day, you reread the page: it, too, turns out to be an accurate description of the day. You continue this process for years; unimaginably, and despite your perverse and whimsical attempts to subvert their accuracy, the books continue to meticulously depict your days.

Here's the question: can you now say, with certainty, that the page that corresponds to today's tomorrow will also be accurate? No! Based on the number of books in the Library and the number of ways in which the description may be inaccurate, despite the long streak of accurate descriptions, it is almost a certainty that the book will deviate. This is a disturbing and counterintuitive conjunction of probability with the comprehensiveness of the Library, yet it is unavoidable. Perhaps this example might help clarify the point. Suppose you flipped a fair coin 15,000 times, which is about once a day for 40 years, and it always came up heads. If you flipped the coin tomorrow, would you expect it to be heads or tails? Of course you'd expect it to be heads again—but if it's a fair coin, there's an equally likely chance it will be tails! A closer correspondence to the probabilities associated with such a book might be: suppose that every day for the past 20 years, you've won the big jackpot of the daily lottery. Do you believe you'll win tomorrow, too? The odds are tremendously against it, but then again, the odds were even more incredible against your winning every day for 20 years. How can you rationally assess tomorrow?

Merrell's book, *Unthinking Thinking: Jorge Luis Borges, Mathematics, and the New Physics*, is the most comprehensive attempt to link ideas of modern mathematics, physics, and philosophy with Borges, via the critical tools of literary analysis. As such, *Unthinking Thinking* contains a number of interesting insights and juxtapositions. For example, Merrell offers unique perspectives on the structure of the Library as seen through the lenses of the theory of special relativity and the expanding universe theory.

I think Merrell makes solid contributions in two areas. I particularly enjoyed his thoughts regarding enantiomorphic (mirror-reversed) forms. He provides a nice discussion of mirror-reversal in the Möbius band and applies his notion imaginatively to the "problem" of mirrors in the Library, especially in reference to his relativistic "world-lines" of librarians.

Second, Merrell gives four arguments for the impossibility of deriving a global order of the Library from the local information that a librarian would have available. Merrell's arguments run the gamut from intertextual references to Borges' story "Averroes' Search" to an appeal to authorities on probability; in particular, Spencer-Brown and the astronomer Layzer, as quoted in Campbell.

Here is a concrete way of thinking about this problem. Suppose I provide you with a rule to generate a sequence, something such as "Start with the two numbers 0, 1. Forever after, employ Rule Fib."

> Rule Fib: *The next number in the sequence is defined to be the sum of the preceding two numbers.*

Rule Fib entails that to find the third number, you must add the first two numbers:

$$0, 1, 0 + 1 = 0, 1, 1.$$

To get the fourth number, you add the second and third terms, $1 + 1$, and get:

$$0, 1, 1, 2.$$

To get the fifth number, you add the third and fourth terms, $1 + 2$, and get:

$$0, 1, 1, 2, 3.$$

The sequence—actually a famous sequence, known as the *Fibonacci sequence*—begins to grow rapidly:

$$0, 1, 1, 2, 3, 5, 8, 13, 21, 34, 55, 89, \ldots$$

Given time and inclination, you or a computer could generate many numbers of this sequence. Conversely, if I provided you with the sequence above, you might well guess the rule that produces "the next term." However, the rule might be much more complicated; for example, it might be: "Let the first 12 terms correspond to the Fibonacci sequence; let the next 12 be the first 12 digits in the decimal expansion of π; the 47 digits after that should all be 7s; etc. etc. etc." Given a more complicated rule such as this last one, although your guess is "good," it relies on the false assumption that the rule generating the sequence must be as simple

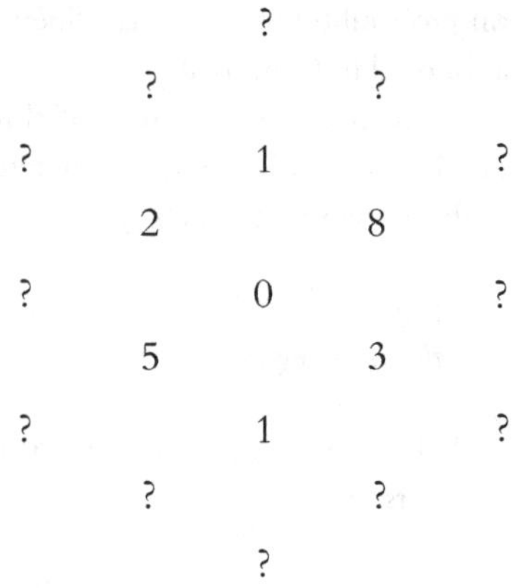

138 ◦∞ UNIMAGINABLE MATHEMATICS

```
              ?
           ?     ?
        ?     1     ?
           2     8
        ?     0     ?
           5     3
        ?     1     ?
           ?     ?
              ?
```

FIGURE 69. The first seven terms of the Fibonacci sequence are distributed in a hexagonal pattern. How should the next 12 terms be placed?

as possible. This has the unfortunate effect that your good guess produces the wrong answer. I conclude that without complete information, there is no way to ensure the successful induction of a *unique* generating-rule for a sequence.

The problem is much worse in higher dimensions. The example of the Fibonacci sequence is one-dimensional; now let's look at one example of the "guessing-of-terms" problem in a two-dimensional setting. The numbers in the hexagonal array pictured in figure 69 are the first seven terms of the Fibonacci sequence, and let's imagine that all the digits to fill out the plane are exactly those appearing in the Fibonacci sequence. But how are the next twelve terms to be ordered? Your guess is as good as mine: There are a vast number of distribution rules utilizing the Fibonacci digits which would produce the pictured hexagonal array. Now imagine a librarian's dilemma, confronting 410-page collections of seemingly random lexical symbols—not even numbers—distributed in some sort of *three*-dimensional lattice. The mind balks at conceiving of any rule to order the books.

A final pair of observations that fit this chapter best. First, Borges introduces the belief of the Book-Man.

> We also have knowledge of another superstition from that period: belief in what was termed the Book-Man. On some shelf in some hexagon, it was argued, there must exist a book

that is the cipher and perfect compendium *of all other books*, and some librarian must have examined that book; this librarian is analogous to a god.

A cipher is either a key or a code, and a compendium is, according to various dictionaries, a brief, a condensation, an epitome, or an abstract. My guess is that Borges meant that since it is possible to conceive of a book 410 pages in length that is a key to and an abridgement of the Library, that book must therefore exist in the Library. Such a book might, in today's parlance, be a computer algorithm for generating all possible symbol sequences of length 1,312,000 from an alphabet of 25 orthographic symbols, for such an algorithm could actually be written in just a few lines of code. More of the Book might be devoted to the generating principles and topology of the Library; possibly a rule for ordering the books (although probably no such rule could fit in one volume); the motivations of the constructors of the Library; how the Library was built and where the materials for it came from; how librarians entered the system; etc. etc. Since such a book can be conceived, the import of Borges' footnote is precisely that it must appear in the Library. Of course, its refutations also exist in the Library, a fact that highlights and compounds the problem of interpretation of truth.

Second, many critics, including some of those mentioned here, have speculated about the meaning and significance of one of Borges' parenthetical asides in the story:

> (Mystics claim that their ecstasies reveal to them a circular chamber containing an enormous circular book with a continuous spine that goes completely around the walls. But their testimony is suspect, their words, obscure. That cyclical book is God.)

I won't presume to provide an exegesis of the cyclical book, but I offer the following insight for a future critic who might wish to interpret it: I believe that again Borges is winking at the reader.

> *It would be impossible to remove such a book from the shelf!*

The only way to read the book would be to physically cut out sections; in other words, the only way for the mystics to attain the Book that is God would be to destroy It. The Book is closed (figure 70).

FIGURE 70. The cyclical Book.

NINE

Openings

To open a book brings profit.

—Chinese proverb

IN THIS CHAPTER, I ASSEMBLE SOME FACTS FOR THE purpose of sketching a picture of the mathematics Borges may have known and how it may have affected the story. In his prologue to the first part of *Ficciones*, Borges winks yet again at the reader when he writes "I am not the first author of the narrative titled 'The Library of Babel'; those curious to know its history and its prehistory may interrogate a certain page of Number 59 of the journal *Sur*, which records the heterogenous names of Leucippus and Lasswitz, of Lewis Carroll and Aristotle." This is precisely the issue of *Sur* in which his essay "The Total Library" appears.[1]

Perhaps few others have had the patience to ferret out the particulars of a hint of Borges' knowledge of combinatorics. Borges opens the story with the following fragment from Burton's *The Anatomy of Melancholy*: "By this art you may contemplate the variation of the 23 letters..." The entire section of Burton is concerned with ways of diverting and amusing oneself, ostensibly towards the end of avoiding or curing melancholy.[2] For several pages before the excerpt, Burton waxes erudite on the pleasures of reading, especially scripture, and of libraries. Without even a paragraph break to ease the transition, Burton moves to pleasures mathematical (emphasis added):

> I would for these causes wish him that is melancholy to use both human and divine authors, voluntarily to impose some task upon himself, to divert his melancholy thoughts: to study the

art of memory, Cosmus Rosselius, Pet. Ravennas, Scenkelius's Detectus, or practise Brachygraphy, &c., that will ask a great deal of attention; or let him demonstrate a proposition in Euclid, in his last five books, extract a square root, or study Algebra; than which, as Clavius holds, "in all human disciplines nothing can be more excellent and pleasant, so abstruse and recondite, so bewitching, so miraculous, so ravishing, so easy withal and full of delight," *omnem humanum captum superare videtur.* By this means you may define *ex ungue leonem*, as the diverb is, by his thumb alone the bigness of Hercules, or the true dimensions of the great Colossus, Solomon's temple, and Domitian's amphitheatre out of a little part. *By this art you may contemplate the variation of the twenty-three letters, which may be so infinitely varied, that the words complicated and deduced thence will not be contained within the compass of the firmament; ten words may be varied 40,320 several ways*; by this art you may examine how many men may stand one by another in the whole superficies of the earth...

It's worth mentioning that the number of distinct ways to order eight words is

$$8! = 40,320.$$

Perhaps Burton had neither the skill nor the stomach to continue multiplying 40,320 by 9 and then again by 10, which would yield 3,628,800, the number of different ways to order 10 words. Whether or not Borges would have recognized this number is moot, yet in his 1936 essay "The Doctrine of Cycles," he correctly calculates the number of ways that the order of 10 atoms can be permuted.

Regardless, he was aware that the passage alluded to combinations and permutations, and that "the words complicated and deduced thence will not be contained within the compass of the firmament." Later in the story, Borges' use of the phrase "the rudiments of combinatory analysis, illustrated with examples of endlessly repeating variations" shows that Borges understood the ideas well, even if a modern mathematician would more likely employ the phrase "variations with unlimited repetition."

Beyond gleaning the story and *Selected Non-Fictions* for clues about his knowledge and predilections, I was fortunate to find another source of information. The chapter title, "Openings," stems from an intersection of optimism and pseudo-randomness. While visiting the National Library of Argentina, I had the great pleasure of perusing the math and science books Borges donated to the collection. I applied the principle that a book beloved by its owner, when held gently underneath the spine and allowed to fall open, will naturally reveal an oft-consulted page. My excitement at achieving interesting results was matched by my chagrin when, after multiple applications of this "opening" principle, I discovered that Borges marked the back end leaves of his volumes with his name, the year of acquisition, and the page numbers—coupled with a succinct phrase—of passages that especially interested him. My chagrin was tempered by the fact that his annotated page numbers unmistakably corresponded with my optimistic openings.

I'll begin with a book that postdates "The Library of Babel," one that evinces that Borges hadn't lost interest in the idea of the Library. In 1949, Borges acquired Russell's *Human Knowledge: Its Scope and Limits*. One of his three annotations on the end leaf is "Eddington's monkeys." Here is the passage from page 484 (emphasis added):

> Eddington used to suggest as a logical possibility that perhaps all the books in the British Museum had been produced accidentally by monkeys playing with typewriters.[3] There are here two kinds of improbability: in the first place some of the books in the British Museum make sense, whereas the monkeys might have been expected to produce only nonsense [...] Suppose you have in your hands two copies of the same book, and suppose you are considering the hypothesis that the identity between them is due to chance: the chance that the first letter in the two books will be the same is one in twenty-six, so is the chance that the second letter will be the same, and so on. *Consequently the chance that all the letters will be the same in two copies of a book of 700,000 letters is the 700,000th power of* $\frac{1}{26}$.

Russell derives a viewpoint complementary to that of the Library. If there are 700,000 letters per book and an alphabet of 26 letters, then the total number of books is $26^{700,000}$. Therefore, the probability of picking a book that exactly matches another is one in $26^{700,000}$; that is

$$\frac{1}{26^{700,000}} = \left(\frac{1}{26}\right)^{700,000}.$$

Many commentators have pointed towards Borges' amiable review of Kasner and Newman's *Mathematics and the Imagination* as an indication of his interest in mathematics and also as a source of his knowledge. Unfortunately, it was not among the books from his personal library that were donated to the National Library. However, I was able to obtain a copy elsewhere and give it a professional reading. (An evocative aspect of the book is that the cover, as opposed to the dust jacket, is embossed with an aleph-nought, $\aleph_0$, which was Cantor's symbol for a countably infinite set.[4]) Borges' review, reprinted in *Selected Non-Fictions*, notes that the book includes

> ... the endless map of Brouwer,[5] the fourth dimension glimpsed by More and which Charles Howard Hinton claims to have intuited, the mildly obscene Möbius strip, the rudiments of the theory of transfinite numbers, the eight paradoxes of Zeno, the parallel lines of Desargues that intersect in infinity, the binary notation Leibniz discovered in the diagrams of the *I Ching*, the beautiful Euclidean demonstration of the stellar infinity of the prime numbers, the problem of the tower of Hanoi, the equivocal or two-pronged syllogism.

Most surprising to me, given that many today attribute an interest in *fractals* to Borges, is that Kasner and Newman's book examines the famous Koch snowflake curve in some depth on pages 344–55. The snowflake curve is a standard introductory example of a fractal—and for historical context, I mention that Kasner and Newman's discussion precedes the term "fractal" by almost 40 years. Apparently, though, Borges was sufficiently unimpressed by the snowflake curve that he neglected to mention it in his review.

Perhaps Borges found the anti-Nazi gibes another appealing facet of the book, given his own strong—and unpopular—anti-Nazi stance during World War II. Despite these many commendable contents and qualities, given that the book was published in 1940, it seems unlikely that it was available for his consultation and degustation prior to the writing of "The Library of Babel."

There are at least two candidates from Borges' personal library to which it is tempting to assign influential status in the development of his mathematical thought. The first is Henri Poincaré's 1908 book

Science et Méthode. Borges' end leaf notations, dated 1939, indicate an interest in Lesage's discredited theory of gravitation and, more tellingly, in geometry and Cantor. One paragraph, taken from pages 380–81, is a passage on geometry worth quoting (emphases added):

> A great advantage of geometry lies in the fact that in it the senses can come to the aid of thought, and help find the path to follow, and many minds prefer to put the problems of analysis into geometric form. Unhappily, our senses can not carry us very far, and they desert us when we wish to soar beyond the classical three dimensions. *Does this mean, beyond the restricted domain wherein they seem to wish to imprison us, we should rely only on pure analysis and that all geometry of more than three dimensions is vain and objectless?* [...] We may also make an *analysis situs* of more than three dimensions. The importance of *analysis situs* is enormous and can not be too much emphasized; the advantage obtained from it by Riemann, one of its chief creators, would suffice to prove this. We must achieve its complete construction in the higher spaces; *then we shall have an instrument which will enable us really to see in hyperspace and supplement our senses.*

Again, I don't imagine that Borges considered exotic cosmologies for the Library, but it interests me to think that he was aware of things living in higher-dimensional spaces.

The sections pertaining to Cantor mainly restrict themselves to exuberant denunciations of set theory via what Poincaré terms "the Cantorian antinomies"—paradoxes arising from Cantor's theory of transfinite numbers. In many ways, Poincaré prefigures a movement towards constructivism in mathematics, which I briefly discuss in the Math Aftermath "Libits, Uniqueness, and Jumping from the Finite to the Infinite." Since Borges was evidently fascinated by transfinite numbers and the concept of infinity, it's striking that as an autodidact, he pursued the arguments and weighed the objections of Poincaré, a formidable opponent of all things infinite.

The other book from Borges' library, philosophically opposed to Poincaré's, is Bertrand Russell's *Principles of Mathematics*. The book was originally published in 1903, and Borges' copy is a 1938 printing. Borges dated his copy "1939," and his annotations further indicate that it was a gift from "Adolfo" (presumably his life-long friend, colleague, and coauthor Bioy Casares). The easiest opening of this volume, and the first page singled out by Borges, concerns a resolution of Parmenides' paradox. The next page pleasantly segues into a discussion of Zeno's paradox of Achilles and the tortoise. The argument contained therein is similar to Russell's refutation in *Mathematical Philosophy*, which Borges outlined in his 1929 essay "The Perpetual Race of Achilles and the Tortoise." Indeed, Borges' annotation includes the phrase "(cf. *Mathematical Philosophy*, 138)."

Borges' essay on Zeno's paradox not only betrays a fondness for and knowledge of Cantor's transfinite numbers; it also demonstrates that Borges understood at least the basics of summing infinite series. That Borges persisted in using Russell as a mathematical touchstone is further evidenced by the brilliant 1939 essay "When Fiction Lives in Fiction," which appears in *Selected Non-Fictions*, pages 160–62. Here, Borges writes

> ... Fourteen or fifteen years later, around 1921, I discovered in one of Russell's works an analogous invention by Josiah Royce, who postulates a map of England drawn on a portion of the territory of England: this map—since it is exact—must contain a map of the map, which must contain a map of the map of the map, and so on to infinity...

Principles of Mathematics is rife with Russell's perspectives on Cantor, transfinite numbers, infinitesimals, the meaning of zero, and a host of other Borgesian obsessions. Despite his cavil found on page 46 of *Selected Non-Fictions* that some of Russell's works are "unsatisfactory, intense books, inhumanly lucid," Borges returned to them again and again. Russell's book, although dry, discursive, and monolithic in conception and execution, contains poetic phrases, one of which Borges singled out with an end leaf notation:

> *... the infinite regress is harmless.*

A point needs to be stressed. Mathematics is a body of lore and an art that requires years of study and practice to understand and appreciate. Just as with twentieth-century atonal music, repeated exposure is required to acculturate the novice to the aesthetics of beauty and elegance particular to mathematics. Grappling with problems and attempting to produce one's own proofs using the licit logical structures are essential to internalize an understanding of the many subtleties inherent in mathematics. I contend that, in this sense, Russell's books are *not* mathematics; rather they are the *philosophy* of mathematics. Therein lay their appeal to Borges, and that is why Russell, not Kasner and Newman, remained Borges' inspiration and touchstone of mathematical thought.

I'll close the book with a last opening: Borges' solitary annotation on the end leaf of Kesten's *Copernicus and his World*. There Borges inscribed a Latin phrase from Copernicus's *De revolutionibus orbium coelestium* and referenced the page containing the English translation.

> *Mathemata mathematicis scribuntur.* "On mathematics, you write for mathematicians only."

It is my hope that this book belies that sentiment.

Appendix—Dissecting the 3-Sphere

> *We sail within a vast sphere, ever drifting in uncertainty, driven from end to end. When we think to attach ourselves to any point and to fasten to it, it wavers and leaves us; and if we follow it, it eludes our grasp, slips past us, and vanishes for ever. Nothing stays for us.*
>
> —Blaise Pascal, *Pensées*

The aim here is to see that three-dimensional slices of a 3-sphere are, in fact, either points or 2-spheres. (We employed this notion in our discussion in the chapter "Topology and Cosmology" when we relied on lower-dimensional analogues to yield insight into the nature of the 3-sphere.) For those whose are interested in this kind of inquiry but whose memory of the equations of spheres and circles is confined to a misty past, we recommend first reading the second section of this appendix, which carefully uses the Pythagorean theorem and the notion of distance in Euclidean space to derive the analytic equations for a circle, 2-sphere, and 3-sphere.

A way to understand three-dimensional slices is to use the analytic equation that defines the unit 3-sphere,

$$w^2 + x^2 + y^2 + z^2 = 1,$$

which should be understood as "the set of all points (w, x, y, z) in coordinatized four-dimensional space that satisfy the above equation." For example the point (1, 0, 0, 0) satisfies the equation, as do the points (0, 1, 0, 0) and (1/2, 1/2, 1/2, 1/2). For the latter point, note that

$$\left(\frac{1}{2}\right)^2 + \left(\frac{1}{2}\right)^2 + \left(\frac{1}{2}\right)^2 + \left(\frac{1}{2}\right)^2 = \frac{1}{4} + \frac{1}{4} + \frac{1}{4} + \frac{1}{4} = 1.$$

If we fix w, the coordinate for the fourth dimension, at 0, the equation becomes

$$0^2 + x^2 + y^2 + z^2 = 1;$$

in other words, the equation of the standard unit 2-sphere. If we fix $w = 1$ (or -1) we are at the top or bottom of the unit 3-sphere, and the equation becomes

$1^2 + x^2 + y^2 + z^2 = 1$, which implies that $x^2 + y^2 + z^2 = 0$.

The only way that three nonnegative numbers can add up to 0 is if they themselves are all 0. In other words, the three-dimensional slice at the coordinate $w = 1$ yields only the point $(x, y, z) = (0, 0, 0)$.

On the other hand, let w be any number strictly between -1 and 1. For a concrete example, let w be 1/2. Then the equation becomes

$\left(\frac{1}{2}\right)^2 + x^2 + y^2 + z^2 = \frac{1}{4} + x^2 + y^2 + z^2 = 1$, which implies that

$$x^2 + y^2 + z^2 = \frac{3}{4}.$$

By taking the square root of both sides, we arrive at

$$\sqrt{x^2 + y^2 + z^2} = \sqrt{\frac{3}{4}} = \frac{\sqrt{3}}{2},$$

and this is equivalent to the statement "the set of all points in three-dimensional Euclidean space located at a distance $\sqrt{3}/2$ from the origin $(0, 0, 0)$." In other words, the equation specifies a 2-sphere of radius $\sqrt{3}/2$.

In the above argument, there was nothing special about letting w be 1/2. We could have chosen any number strictly between -1 and 1, and we would again end up with an equation specifying a sphere.

In general, let $w = R$, where $-1 < R < 1$. Then the examination of the three dimensional slice at $w = R$ is facilitated by the equation

$$R^2 + x^2 + y^2 + z^2 = 1, \text{ which implies that}$$
$$x^2 + y^2 + z^2 = 1 - R^2.$$

By taking the square root of both sides, we arrive at

$$\sqrt{x^2 + y^2 + z^2} = \sqrt{1 - R^2},$$

which is the equation for a 2-sphere of radius $\sqrt{1 - R^2}$.

Deriving the Equations for Circles and Spheres Via the Pythagorean Theorem

The Pythagorean theorem states that for a right triangle with legs of lengths x and y and with hypotenuse of length h contained in a Euclidean plane, the equation $x^2 + y^2 = h^2$ always holds (figure 71). (There are dozens, maybe hundreds, of proofs of this theorem.) If the length of the hypotenuse is, say, $1/2$, then the equation becomes

$$x^2 + y^2 = \left(\frac{1}{2}\right)^2,$$

and taking the square root of both sides of the equation yields

$$\sqrt{x^2 + y^2} = \frac{1}{2}.$$

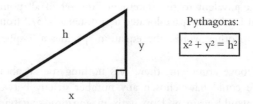

FIGURE 71. An illustration of the Pythagorean theorem.

In fact, it follows instantly that if h is the length of the hypotenuse, then Pythagoras implies

$$\sqrt{x^2 + y^2} = h.$$

Thus the key point is the realization that the length of the hypotenuse is expressible in this form. Now, think of the bottom-left point of the hypotenuse as the origin $(0, 0)$ of the plane and reimagine the lengths of the legs, x and y, as representing the horizontal and vertical coordinates for the right-top point of the hypotenuse. Then the length of the hypotenuse signifies the distance from the origin to the point $p = (x, y)$, and applying the Pythagorean theorem reveals the distance to be $\sqrt{x^2 + y^2}$. See figure 72.

Now, we want to use these ideas in order to derive an equation equivalent to Euclid's intuitively satisfying definition of a circle. He defined a circle to be the set of points in a plane that are equidistant from a given point. If we set the given point to be the origin, and choose the distance to be equal to one, then a circle is the set of all points (x, y) that satisfy the equation.

$$\sqrt{x^2 + y^2} = 1.$$

After squaring both sides, we see that it must be the case that a unit circle is precisely all points (x, y) that fit this equation:

$$x^2 + y^2 = 1^2 = 1.$$

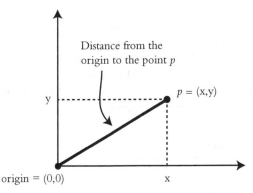

FIGURE 72. Using Pythagoras to define distance in the Euclidean plane.

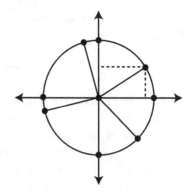

FIGURE 73. Using the notion of distance, hence Pythagoras, to define a circle.

This is how the analytic equation for the circle arises, and figure 73 indicates a way of viewing a circle as a composition of distances from the origin, that is, as hypotenuses of right triangles.

The equation for a 2-sphere is very similar in concept, and thus we need only adapt our notion of distance—and therefore, the Pythagorean theorem—to work in three-dimensional space. A standard way is a typically incisive mathematical maneuver which requires the clever use of the Pythagorean theorem twice.

To see this, let's find the distance from the origin $(0, 0, 0)$ to the point $p = (x, y, z)$ in coordinatized 3-space. The point p naturally determines a right triangle, with the first leg of the triangle being the line segment contained in the $x - y$ plane (for which $z = 0$) that connects the origin to the point $(x, y, 0)$. The second leg is the vertical line segment connecting the points p and $(x, y, 0)$. The hypotenuse of this right triangle is the distance we want—see figure 74.

Observe that the length of the leg that connects p to the point $(x, y, 0)$ is simply the height, z. Since the other leg is contained in the x-y plane for which z is constantly 0, at a critical juncture below we will ignore the z coordinate and blithely apply the Pythagorean theorem as we did above in the Euclidean plane. First, though, using the Pythagorean theorem on the dark gray triangle in figure 74 gives

$$\underbrace{\left(\text{distance from origin to } (x, y, 0)\right)^2}_{\text{first leg}} + \underbrace{(\text{height } z)^2}_{\text{second leg}}$$

$$= \underbrace{\left(\text{distance from origin to } (x, y, z)\right)^2}_{\text{hypotenuse}}$$

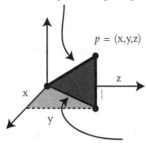

Use Pythagoras in the plane to find the hypotenuse of the light gray triangle. Now, use this hypotenuse, along with the distance z, to find the hypotenuse of the dark gray triangle!

FIGURE 74. Using the Pythagorean theorem twice to define distance in 3-space.

Again, because the second leg of the dark gray triangle is the hypotenuse of the light gray triangle in (essentially) the x-y plane, the Pythagorean theorem allows us to replace "distance from origin to $(x, y, 0)$" with "$\sqrt{x^2 + y^2}$." We also may think of "height z" as just "z," and making these substitutions transforms the previous equation into

$$\left(\sqrt{x^2 + y^2}\right)^2 + (z)^2 = \left(\text{distance from origin to } (x, y, z)\right)^2.$$

Squaring the square root in the first term of the above equation and dropping the parentheses leaves us with

$$x^2 + y^2 + z^2 = \left(\text{distance from origin to } (x, y, z)\right)^2.$$

Taking the square root of both sides of the equation yields

$$\sqrt{x^2 + y^2 + z^2} = \text{distance from origin to } (x, y, z).$$

One may now analytically define the unit 2-sphere in the same way the circle was defined; it is the set of points (x, y, z) contained in 3-space that all are of distance one from the origin. This translates into the fact that the 2-sphere is the set of all points (x, y, z) that satisfy the distance equation

$$\sqrt{x^2 + y^2 + z^2} = 1.$$

And thus, by squaring both sides, we arrive the analytic equation for the 2-sphere:

$$x^2 + y^2 + z^2 = 1^2 = 1.$$

Generalizing these ideas to coordinatized four-dimensional Euclidean space is similar—we need only adapt our notion of distance to 4-space. We do this by again bootstrapping ourselves into a higher dimension by cleverly using Pythagoras twice.

Let $p = (w, x, y, z)$ be a point in 4-space—and notice this time that the "new" coordinate is added in front of, rather than behind, the previous coordinates. Once again, the point p naturally determines a right triangle in 4-space (which unfortunately we are unable to draw) with the first leg of the triangle being the line segment connecting the origin to the point $(0, x, y, z)$, and this segment is completely contained in the x-y-z Euclidean 3-space for which the w coordinate is constantly equal to 0. The second leg of the triangle is the line segment "vertically" connecting (w, x, y, z) to $(0, x, y, z)$; in other words, a leg of "height" equal to w. The hypotenuse of the triangle connects the origin of 4-space to the point p, and is the distance we want. So applying Pythagoras to the right triangle yields

$$\underbrace{(\text{height } w)^2}_{\text{first leg}} + \underbrace{\left(\text{distance from origin to } (0, x, y, z)\right)^2}_{\text{second leg}}$$
$$= \underbrace{\left(\text{distance from origin to } (w, x, y, z)\right)^2}_{\text{hypotenuse}}$$

The distance formula in three dimensions derived earlier allows us to replace "distance from origin to $(0, x, y, z)$" with "$\sqrt{x^2 + y^2 + z^2}$," and "height w" is simply equal to "w." Making these substitutions transforms the equation into:

$$(w)^2 + \left(\sqrt{x^2 + y^2 + z^2}\right)^2 = \left(\text{distance from origin to } (w, x, y, z)\right)^2.$$

Again, squaring the square root in the first term of the above equation and dropping the parentheses leaves us with

$$w^2 + x^2 + y^2 + z^2 = \left(\text{distance from origin to } (w, x, y, z)\right)^2.$$

Taking the square root of both sides yields

$$\sqrt{w^2 + x^2 + y^2 + z^2} = \text{distance from origin to } (w, x, y, z).$$

The unit 3-sphere is the set of all points in 4-space uniformly a distance one from the origin. This is equivalent to the set of all points (w, x, y, z) satisfying the distance equation

$$\sqrt{w^2 + x^2 + y^2 + z^2} = 1,$$

and, after squaring both sides, we end up with

$$w^2 + x^2 + y^2 + z^2 = 1^2 = 1.$$

Notations

> *The twentieth-century artist who uses symbols is alienated because the system of symbols is a private one. After you have dealt with the symbols you are still private, you are still lonely, because you are not sure anyone will understand it except yourself.*
> —Louise Bourgeois, quoted in *Lives and Works*

$a \approx b$ This says that for our intents and purposes, we can think of a as being roughly equivalent to b: "a is approximately b."

$a \times b$ One way of signifying the product of a with b; that is, a way of notating the act of multiplying a by b.

$a \cdot b$ Another way of signifying the product of a with b; that is, another way of notating the act of multiplying a by b.

$n!$ Read "n factorial," and
$$n! = n \cdot (n-1) \cdot (n-2) \cdot (n-3) \cdot \cdots \cdot 3 \cdot 2 \cdot 1.$$

a^b The integer a raised to the bth power, for example, $3^5 = 3 \cdot 3 \cdot 3 \cdot 3 \cdot 3 = 729$.

$[a, b]$ The closed interval between a and b. All numbers between a and b, inclusive.

∞ A symbol for infinity. It's important to note that although ∞ connotes a kind of magnitude, and is sometimes shorthand for the idea of "arbitrarily large number," the symbol ∞ is not a number.

Notes

The end crowneth the work.
    —Elizabeth I, quoted in *The Sayings of Queen Elizabeth*

The end crowns all;
And that old common arbitrator, Time,
Will one day end it.
    —William Shakespeare, *Troilus and Cressida*

Preface

1. Did you look?

Chapter 1

1. For example, Lasswitz, who wrote "The Universal Library," which profoundly influenced Borges, calculated the number of books in his Library. Other mathematicians and critics who find the number of books include Amaral, Bell-Villada, Rucker, Nicolas, Faucher, Salpeter, and the anonymous encyclopediasts who wrote the page found at Wikipedia.org! Amaral deserves special plaudits for finding influences of Lasswitz's "The Universal Library" in the work of Lasswitz's mathematical contemporaries Kummer, Fraenkel, pp. 7ff, and Hausdorff, pp. 61ff.
2. The quote below appears in Borges' expansive short story "Tlön, Uqbar, Orbis Tertius."

   There are no nouns in the conjectural *Ursprache* of Tlön, from which its "present-day" languages and dialects derive: there are impersonal verbs, modified by monosyllabic suffixes (or prefixes) functioning as adverbs. For example, there is no noun that corresponds to our word "moon," but there is a verb which in English would be "to moonate" or "to enmoon." "The moon rose above the river" is "*hlör u fang axaxaxas mlö*," or, as Xul Solar succinctly translates: *Upward, behind the outstreaming it mooned.*

His use of the phrase "*Axaxaxas Mlö*" in "The Library of Babel" is presumably a reminder that even books written in the *Ursprache* of Tlön, including all volumes of the first and second editions of the *Encyclopedia of Tlön*, are in the Library. A careful reader may object that volume 11 of the *First Encyclopedia of Tlön* consists of 1,001 pages, while Library books number only 410. Our rejoinder is that three books of the Library, the last of which will contain 229 blank pages—blank spaces filling each slot—yield the necessary 1,001 pages. Of course, they may not be shelved anywhere near each other, but this in no way negates the fact that the 11th volume *is* in the Library. A variation of this observation refutes Rucker's casual statement that "the minute history of the future" can't be contained in the Library, in *Infinity and the Mind*, pp. 121–22. The minute history of the future is contained in the Library; it is found in volumes perhaps scattered throughout the Library. There is no implicit promise that the information in the Library is accessible or verifiable—it just must be there, somewhere.

3. From the 1875 Grindon citation under the definition of "septillion" in the *Oxford English Dictionary*. "Thousands of plants consist of nothing more than a few such cells as in septillions make up an oak-tree..."

4. In the words that comprise "The Library of Babel," Borges adroitly finesses the fact that by being restricted to a mere 25 symbols, the books of the Library could not contain both upper- and lower-case symbols, let alone diacritical and punctuation marks beyond the comma and the period. The story, as written, could not appear in the Library. In this book, when we are referring to imagined entries in the Library, we will hew to the standard set by Borges and not restrict ourselves to using only the orthographic symbols of all upper-case letters (or lower-case letters), spaces, periods, and commas.

Chapter 2

1. For a mathematically sophisticated reader: in fact, one may imagine that at some distant point in the future, we might have a superfast supercomputer running an algorithm that

    1. Was able to test for primality every number expressible in 100 digits.
    2. Kept a tally of the number of primes without listing them.
    3. Output simply the number of primes N expressible in 100 digits.

    Then,

    4. Determine if N is odd or even.
    5. Determine, if N is odd, which numbered prime is the median of the set.

6. Determine, if N is even, which two numbered primes average to the median of the set.

Then,

7. Run the algorithm again, keeping count until the number(s) from step 5 (or 6) is (are) achieved.
8. Output the median!

At no point was a list of all primes necessary.

Chapter 3

1. See Salpeter, for example, on "The Book of Sand."
2. Benardete, quoted in Merrell on page 58, independently thinks along similar lines.
3. Another fine point for the interested: obviously we're capitalizing on the fact that the number of pages is countably infinite.
4. If a mathematically sophisticated reader is worried about the need to invoke the axiom of choice, the issue is easily sidestepped by assigning the same infinitesimal, ß, for the thickness of each page.

Chapter 4

1. For the mathematically adventurous reader: in fact, the famous Hopf fibration of the 3-sphere decomposes the 3-sphere into great circles over a base space equivalent to the 2-sphere.
2. A point for a mathematically sophisticated reader: earlier in this chapter, we observed that the gravitational field of the Library needed to be imposed by the builders of the Library. Since the Library presumably does not possess any regions of zero gravity, it is vital that the 3-manifolds under consideration may be equipped with everywhere nonzero vector fields. But of course they can, since all 3-manifolds have Euler characteristic equal to 0, entailing the existence of everywhere nonzero vector fields.
3. A way to see that the surface of a coffee mug is the same as the surface of a donut is to simply shrink the "cup" part of the mug to the strip that lies between the joining spots of the handle to the mug!
4. Another way around this problem would be to require that the orthographic symbols be symmetric under 90° rotations, too. For example, the symbols O, X, +, ⊗, ⊕, and □ satisfy this criterion.
5. Technical note: as the Library stands, we couldn't actually use an exact cube to construct the Library, for the number of hexagons is not a perfect cube. That is, the number of hexagons is not of the form x^3 for some integer x. However, by tinkering with the number of hexagons on each floor, it would

be possible to have a shape very close to a cube. If done carefully, it would be equally easy to make the identifications between the sides of the near-cube, and the end result would be just slightly less symmetric.

Chapter 7

1. An Argentine colleague, Martin Hadin, brought to my attention a 1971 dialogue between Borges and Herbert Simon that suggests Borges was intrigued by these kinds of ideas, but previously unaware of them. The dialogue appears in *Primera Plana*, January 5, 1971.

Chapter 8

1. The interesting readings in Wheeler, Alazraki, Bell-Villada, Barrenechea, Rodriguez Monegal, Slusser, Ammon, Eco, Keiser, Nicolas, and Faucher, for example, fall outside this domain. Although I might differ with the conclusions they draw, it seems to me that Nicolas and Faucher get the math correct.
2. An infinite set is *countable* (also called *denumerable* or more precisely, countably infinite) if it can be placed into one-to-one correspondence with the positive integers. In effect, this means that one may write down all the elements of the set in an orderly (infinitely long) list.

   1. ↔ "first" element of the set
   2. ↔ "second" element of the set
   3. ↔ "third" element of the set
   etc.

   It's easy and not inaccurate, therefore, to think of "countable" as synonymous with "listable." Cantor, the creator of set theory and the theory of transfinite numbers, was presumably shocked to discover that the *rational numbers* are countable. Regardless, in one of the most beautiful arguments in mathematics, he demonstrated that the *irrational numbers* are *not* countable or listable. Any set that is not listable is called *uncountable* or uncountably infinite.
3. If, for example, I begin by assuming, "The moon is made of green cheese," I can derive a whole host of interesting implications from that premise, such as that dairy farmers and cheese manufacturers may be responsible for the lack of missions to the moon in recent years. That if we built cheese-harvesting factories on the moon and sent back enormous loads of cheese, the world hunger situation might be abated. And so on. However, the original premise is false, so it doesn't matter how interesting or plausible the ensuing speculations are to us.

Chapter 9

1. For the reference to "The Total Library," see *Collected Fictions*, p. 67. For the essay itself, see *Selected Non-Fictions*, p. 214.

2. See Burton, p. 95. The quote is lifted from Part 2, Sec. 2, Memb. 4 of Burton's colossus.
3. A marvelous, witty experiment was performed at the Paignton Zoo in Devon in 2003. Six Sulawesi crested macaque monkeys were placed in a cage outfitted with a computer, ostensibly to see if, between them, the monkeys might produce some work of Shakespeare over the course of a week of random typing. As reported by David Adam, science correspondent for *The Guardian*, "The macaques—Elmo, Gum, Heather, Holly, Mistletoe, and Rowan—produced just five pages of text between them, primarily filled with the letter S. There were greater signs of creativity towards the end, with the letters A, J, L and M making fleeting appearances, but they wrote nothing even close to a word of human language."
4. Ferrero and Palacios, Hayles, and especially Hernández have written more about Borges and $\aleph_0$.
5. This "endless map of Brouwer" goes by the mathematical name of Brouwer's fixed point theorem. The main idea is that if a nice enough space is mapped into or onto itself, then there must be at least one point that the map does not move, a fixed point. An intuitive way of seeing this is fundamentally similar to Josiah Royce's construction, which is briefly quoted and discussed on page 146. For Royce, an exact smaller image of England is on a map. But if the map is exact, then there is an unimaginably smaller version of the map on the map. And that version must also have a smaller version contained within. These images, each one contained in the previous, appear to shrink to a point. In fact, they do: the math capturing Royce's idea was formally stated and proved by Banach and others in the early 1920s, and today the result goes by the name of the contraction mapping principle. In *Variaciones Borges*, John Durham Peters takes a deep look at philosophic connections between Borges, Royce, and William James. He also considers Royce-type maps from many perspectives, delineating some interesting mathematical and situational implications, and using these ideas to meditate on the real world in contrast to mathematics.

Glossary

Thus we may define the real as that whose characters are independent of what anybody may think them to be.
—Charles Sanders Peirce, *How to Make Our Ideas Clear*

You who read me—are you certain you understand my language?
—Jorge Luis Borges, *The Library of Babel*

In some cases, the phrases and clauses that follow should be considered more as gestures and less as definitions: they're not necessarily meant to be rigorous or precise, but rather to evoke a way of understanding the mathematical object. For those with internet access and inclination, Wikipedia at www.wikipedia.org stands up as a surprisingly good source for formal definitions. Wolfram's MathWorld at mathworld.wolfram.com is equally good, and perhaps less subject to malicious or mischievous hacks and prankings.

Beyond standing the test of time and invoking chills of the mythologic, the stacks of libraries stocked with math books are invested with the pregnant allure of opening crisp new or musty old books, and then using indices to seek out appearances of the term or concept. By so doing, you may follow Borges' footsteps through dim-lit libraries, tracking the spoor left by the intellectual history of an idea and slowly netting it with your growing framework of context and insight. Libraries are cultural resources eroding byte by byte under the rising tide of digitization. I point this out partly as a lament, but mostly as a tedious reminder for those so inclined to seize the opportunity to use libraries before they change beyond recognition.

GLOSSARY

1-*space* The Euclidean line. The real number line. One-dimensional space.

1-*sphere* A circle contained in a plane. All the points in a plane that are the same fixed distance from a particular point.

2-*space* The Euclidean plane. The Cartesian coordinate plane. Length by width. The x-y plane. Two-dimensional space.

2-*sphere* A basketball. A soccer ball. The generalization of the 1-sphere to a higher dimension. All the points in 3-space that are the same fixed distance from a particular point.

3-*Klein bottle* A three-dimensional analogue of the Klein bottle. A nonorientable object living in higher dimensions that is formed by identifying the faces of a solid cube or hexagonal prism. A somewhat improbable model for the universe that is the Library.

3-*space* The space we appear to live in. Volume. Length by width by height. The x-y-z space. Three-dimensional space.

3-*sphere* The generalization of the 2-sphere to a higher dimension. A geometric object that lives naturally in 4-space. All the points in 4-space that are the same fixed distance from a particular point. A model for the universe that is the Library which satisfies the particulars of the Librarian's classic dictum as well as those of the Librarian's solution.

3-*torus* The generalization of the torus to higher dimensions. A geometrically flat object that lives most naturally in 6-space, although it may inhabit 4-space. A solid object living in higher dimensions that is formed by identifying the faces of a solid cube or hexagonal prism. The most sensible model (whatever that means) for the universe that is the Library.

4-*space* In the context of our universe, 4-space is often called the space-time continuum, and can be thought of as (Volume) × (One time dimension). In this book, though, it's (Volume) × (Another Euclidean dimension). The w-x-y-z space. Four-dimensional space.

annulus An annulus is the area between two concentric circles in the Euclidean plane. Topologically, it is the same as a cylinder, or a can that has had the top and bottom removed.

Archimedean property Often stated in the form that there is no largest integer. This is then usually flipped, by taking reciprocals, to conclude that there is no smallest positive number. It's then easily generalized to point out that all real numbers are beset and besieged by other real numbers, none of which is "closest."

axiom A statement so fundamentally in accord with our intuition and experience of the world that we are willing to accept it as a basis for all future developments. A logical "given."

base of an exponent The number that is multiplying itself some fixed number of times. For example, in the expression $5^3 = 5 \cdot 5 \cdot 5 = 125$, the number 5 is the base of the exponent.

Brouwer's fixed point theorem In its simplest form, Brouwer's fixed point theorems says that if we take any closed disk in the plane and twist it, stretch it, contract it, rotate it, and do what we will with it, and then squish the transmogrified disk back down into the plane so that it lies within its original boundaries, then there must be at least one point that is unmoved. That is, despite all the distortions and contortions, there must be a fixed point.

cardinality The number of elements in a set.

Cavalieri's principle Cavalieri's principle is a way to think of the volume of an object as the sum of infinitely many infinitesimally thin slices of the object. In calculus terms, for a sufficiently "nice" object, we can integrate the areas of the slices to find the volume of the object.

chiliagon A thousand-sided polygon. A chiliagon on the page of this book would be virtually indistinguishable from a circle. Descartes used it as an example of a geometric object that's easy to define but impossible to visually imagine within the mind's eye.

circular logic See "illegitimate deduction."

circumference The circumference of an n-sphere, for any dimension n, is the distance around the equator of the sphere. The distance around any great circle of the sphere. If the radius of the n-sphere is r, then $2\pi r$ is the circumference.

closed interval A closed interval of the real number line is the set of all points between two numbers, inclusive of the endpoints. For example, the closed interval between 1 and 7 is the set of all numbers x such that $1 \leq x \leq 7$.

codimension The codimension of a geometric object living in some n-space is the difference in dimensions between the object and the ambient space. For example, a line is a one-dimensional object. The codimension of the line in 2-space is equal to 1. The codimension of the line in 3-space is equal to 2. The codimension of the line in 4-space is equal to 3. The codimension of the line in n-space is equal to $(n-1)$.

combinatorics Combinatorics is the art of counting something in two different ways, setting those equal to each other, and thereby finding a formula with general applicability. A lot of interesting combinatorics can be done by thinking carefully about the many ways different colored balls can be placed into barrels.

countable A set is countable if it can be put into one-to-one correspondence with the positive integers. If the elements of a countable set are playing musical chairs and there is a chair for each positive integer, when the music stops every element will be able to find a seat, every time. Such a set is also called *countably infinite*.

definition See "definition" or "self-referential."

denominator The denominator of a fraction is the number dividing into the numerator. The bottom of the fraction. The basement of the fraction. In the expression 3/5, the denominator is 5.

empty set, complete list of elements contained within See page xlii.

Euclidean space A space satisfying Euclid's postulates. See also 1-space, 2-space, 3-space, 4-space, etc.

exponential notation A remarkably condensed and useful notation that captures the idea of a number multiplying itself some specified number of times. In this book, we use it only for integer self-multiplications, but the ideas can be extended so that all real numbers are legitimate exponents. With somewhat more difficulty, the ideas may be further extended so that imaginary and complex numbers may also serve as exponents.

factor (noun) A positive integer which, when multiplied by another positive integer, produces yet another positive integer of situational interest.

factor (verb) Given a positive integer, it's the finding of the factors (noun) that multiply each other to produce the original integer.

factorial A useful notation for positive integers that often crops up in combinatorial formulas. Generalized by the gamma function. Approximated by Stirling's formula. See the section "Notations" for a formal definition and an example.

Fibonacci sequence Counts, for each successive generation, the number of immortal rabbits living in an infinite universe. Has terms whose ratios converge to the Golden Mean. Arises in surprising places in nature. Is related to logarithmic spirals. Is the object of study of entire books.

fixed point A fixed point of a function that maps from a space to itself doesn't move. For example, if we map the real numbers to themselves by the function f(x) = 3x, then f(0) = 3·0 = 0, entailing that 0 is a point fixed by the function.

flat Describes a geometric object equipped with a notion of distance which may be precisely the same as in Euclidean space. For example, although curved, the surface of a cylinder is flat, because to find the distance between two points, the cylinder may be "cut open and unrolled" and "laid flat." Then the two points may be connected by a Euclidean straight line and then "rerolled."

fractal An object with a dimension that is not an integer. An object that continues to present visual complexity under increased magnification. Clouds, bark, lungs, leaves, coastlines, strange attractors, the Koch snowflake curve, the Cantor set, ...

function A way of relating two spaces. A way of relating a space with itself. A way of corresponding elements of one set with elements of another set. A systematic process that inputs real numbers and outputs real numbers.

Funes-like Ireneo Funes is a character in a remarkable Borges short story who is gifted and afflicted with essentially perfect memory. Funes spends more than a day reliving every detail of a day.

gamma function An elegant way of generalizing the concept of the factorial of a positive integer to that of all real numbers.

glossary See "definition" or "self-referential."

great circle An equator of an n-sphere. A circle of maximal size that can be contained in an n-sphere.

hexagonal prism A hexagon is a symmetric six-sided object contained in a plane. A hexagonal prism is a three-dimensional object whose horizontal slices are filled-in hexagons.

homomorphism A function maps elements of one set to elements of another set. A homomorphism is a function that also preserves algebraic relations during the mapping; for example, we can think of integers as points in the set of real numbers, but we can also think of numbers as things that do algebraic stuff, such as addition and subtraction. A homomorphism maps integers both as elements and as algebraic objects. If this intrigues, see Gallian's *Contemporary Abstract Algebra*.

hyperreal number An infinitesimal affiliated with any nonzero real number.

hyperreal number line   The real number line, combined with all the hyperreals affiliated with each real number.

illegitimate deduction   See "circular logic."

infinitesimal   An idea used by Euler, Newton, and Leibniz in thinking about calculus. Infinitesimals may be thought of as actual numbers in a logically consistent way, and may loosely be thought of as entities that have "magnitude" greater than 0 but are smaller than every positive real number. Every real number x can logically be thought of as being surrounded by infinitesimal hyperreal numbers that are closer to x than any other real number.

initial position   The starting point for a Turing machine.

integers   The set of whole numbers $\{\ldots -2, -1, 0, 1, 2, \ldots\}$.

internal state   A particular set of instructions for a Turing machine, telling the Turing machine what should be done in response to each possible input.

irrational numbers   The set of real numbers that can't be written in the form of a fraction p/q, where p and q are both integers. When an irrational number is written out in decimal form, the digits in the expansion neither terminate nor turn into a repeating pattern.

Klein bottle   A torus that has lost it's way in 4-space. A boundaryless nonorientable two-dimensional object.

Koch snowflake curve   A pleasantly symmetric example of a fractal that appears in a math book that Borges reviewed. One unusual property that it possesses is that the distance between *any* two points is infinite. The more closely we look at any portion of the snowflake curve, the more detail emerges.

lemma   A lemma is an assertion not quite important enough to be called a theorem.

libit   Short for "library unit." A library unit is a collection of contiguous hexagons sufficiently large to hold all $25^{1,312,000}$ distinct volumes and sufficiently symmetric that copies of it are able to tile the infinite 3-space model of the Library.

locally Euclidean   A space is locally Euclidean if at every point of the space, a severely myopic individual is convinced that they are, in fact, inhabiting a Euclidean space. As an example, consider the circle. It is clearly not a Euclidean line, but if you have access to a math program or drawing program that allows you to zoom in on an object, and you zoom in on any point, you'll find that what began as looking like a curve looks a lot like a straight line. Thus, a circle is locally Euclidean.

logarithm The logarithm is a function characterized by several remarkably useful properties. All of these stem from the fact that it is the inverse function to the exponential function in base 10. (An inverse function cancels the effect of its corresponding function.)

lower bound A minimal estimate. "At least thus-and-such."

manifold A shorter name for a locally Euclidean space.

map Another name for a function, for we can think of a function not only as taking inputs and returning outputs but also as taking a point and mapping it, or moving it, to another point.

median The median of a finite set of numeric data is a kind of a middle number: half of the data will be larger than the median, and half will be smaller.

Möbius band A Möbius band is a nonorientable, one-sided surface with one boundary circle. Taking two Möbius bands and gluing them together along their boundary circles produces a Klein bottle! (This is not an obvious construction.)

non-Euclidean Any space that is not a Euclidean space. For example, all manifolds, including spheres, tori, and Klein bottles, are non-Euclidean. A cylinder, a figure eight, and a spiral are all non-Euclidean. Typically, though, we'd only refer to a space as non-Euclidean if it's everywhere locally Euclidean.

nonorientable space Best defined in opposition to an orientable space: in an orientable two-dimensional manifold, at any point we may choose a definition of "up" and "right," and then, regardless of the path we navigate through the space, when we return to our beginning point our notions of "up" and "right" will agree with those that we originally chose. By contrast, in a nonorientable two-dimensional space, after making choices of "up" and "right," there are circuitous paths we may follow such that when we return to the starting point, either "up" will look like "down" or "right" will appear "left." In a three-dimensional manifold, we'd also have to choose a "front" to make a legitimate definition.

nonstandard analysis Logically sound mathematics done with infinitesimals and hyperreal numbers.

numerator   The numerator of a fraction is the number being divided by the denominator. The top of the fraction. The attic of the fraction. In the expression 3/5, the numerator is 3.

one-to-one correspondence   This is a map between sets A and B such that every element in A is sent to a distinct element in B and every element in B has exactly one element of A mapped to it. If A and B are finite sets, it means that they each have the same number of elements. If A and B are infinite sets, the implication is that they have the same cardinality.

origin   The point in coordinatized n-space such that all coordinates are 0. The point where the axes all intersect. An arbitrarily chosen point that serves as the center of the space.

periodic   A pattern is periodic if it repeats over and over. For example, the pattern of letters MCVMCVMCVMCV is periodic of period 3. The earth orbiting the sun is an example of periodic motion. A wallpaper pattern may be periodic.

power of 10   An exponential expression with a base of 10. Examples include 10^3, 10^{100}, and, more abstractly, 10^n, which signifies "some power of ten."

prime number   A number p whose factors are limited to 1 and p. No other positive integer may divide a prime number.

product   Another name for the act of multiplication.

raised to a power   Another phrase for raising a base by an exponent. Another way of saying that a number is being multiplied by itself a specified number of times.

rational numbers   All numbers of the form p/q, where p and $q \neq 0$ are both integers.

real numbers   The set of all rational and irrational numbers.

real number line   Euclidean 1-space. The set of real numbers identified with points on the Euclidean line.

self-referential   See "self-referential."

set   A collection of objects, usually called elements. If memory serves correctly, "set" is the word in the English language with the most definitions—at any rate, the 2nd edition of the Oxford English Dictionary runs to 23 pages of definitions and citations for the word "set."

set of measure 0 An inconsequential set. A set that essentially occupies none of the ambient space that it lives in. Pick an arbitrarily small number c: a set of measure 0 can be covered by—contained in—countably many sets whose diameters sum to a number smaller than c. (The *diameter* of a set may be thought of as the maximal distance across it.)

set theory One of the foundations of modern mathematics. One of the underlying languages of modern mathematics. A collection of seemingly unassailable intuitions about objects in our world.

space A collection of points, often equipped with some notion of distance between points.

Stirling's approximation to the factorial A way of approximating $n!$ using Euler's constant e, exponentials, square roots, and 2π. See, for example, page 616 of Apostol's *Calculus, Volume II* for a derivation of the formula.

subset A subcollection of a set. A subset of a set can be the whole set, some of the set, or none of the set. The subset consisting of no elements is called the *empty set*. See "empty set."

tiling of space An object tiles a space if clones of the object completely fill out the space with no interstices or overlaps. For example, it's not too hard to see that squares tile the plane and that cubes tile 3-space. Bisecting the squares along diagonals shows that triangles also tile the plane. Looking at a beehive suggests how hexagons may tile the plane, which in turn suggests the correct belief that hexagonal prisms tile 3-space.

topology Very loosely, topology is the study of the possibilities and immutable characteristics of spaces.

torus The surface of a donut or bagel. A good example of a two-dimensional locally Euclidean space.

transfinite numbers These days, most mathematicians would call transfinite numbers either infinite cardinal numbers or infinite ordinal numbers. As the name suggests, transfinite numbers are beyond the finite, and they are truly unimaginable.

uncountable Describes an infinity infinitely larger than countably infinite. The cardinality of the set of irrational numbers.

unique factorization The property enjoyed by the positive integers that they may be decomposed into products of primes in essentially only one way.

upper bound   A maximal estimate. "There are at most thus-and-such."

Venn diagram   A way of viewing unions, intersections, and subsets of collections of sets by representing the sets as circles.

well-ordering principle   A surprisingly contentious axiom or theorem (depending on the system) that says every set of positive integers contains a least element. The reason some mathematicians and logicians reject the well-ordering principle is that it is used to facilitate kinds of deductions that may lead to disturbing conclusions.

Annotated Suggested Readings

> *All books are divisible into two classes, the books of the hour, and the books of all time.*
> —John Ruskin, *Sesame and Lilies*

> *I was impressed for the ten thousandth time by the fact that literature illuminates life only for those to whom books are a necessity. Books are unconvertible assets, to be passed on only to those who possess them already.*
> —Anthony Powell, *The Valley of Bones*

In this section, I list a few readings that in one way or another go deeper into ideas raised in this book. I've loosely organized them, mostly by the chapter that they illuminate. Like most of my book, the list is somewhat idiosyncratic; books and articles appearing tend to have had a lasting impact on me, or, in a few cases, received a strong recommendation from someone I respect. For more personalized recommendations, feel free to write me at <babel.librarian@gmail.com> describing your math background and the kinds of things you'd like to learn. Publication details for each book may be found in the Bibliography.

Generally Delightful

The Heart of Mathematics, by Edward B. Burger and Michael Starbird.

Burger and Starbird produced a funny, inspirational, eminently readable book pitched at the level of bright high school students and college students who haven't (yet) had a lot of training in mathematics. It's almost as if they thought, "What are the niftiest ideas in math that don't need a deep theoretic background? How can we get them all into one book?" and then went ahead and did it. Great problems are found at the end of every chapter, and some answers are included. It's worth mentioning that Starbird is a raconteur of the first order, and Burger worked as a stand-up comedian before becoming a mathematician.

The Pleasures of Counting, by T. W. Körner.

This remarkable book unites a host of topics by the common theme of mathematics making a difference in solving real-world problems. Körner opens the book with a discussion of how Dr. John Snow essentially invented epidemiology when he analyzed data pertaining to cholera outbreaks in the middle 1800s. Körner moves with ease from there through contributions to thwarting submarine warfare, development of radar, cracking the Enigma code, and a host of other fascinating applications.

Generally Thoughtful

"What Is Good Mathematics?" by Terence Tao (*Bulletin of the American Mathematical Society* 44(2007): 623–34, available as a free .pdf download from the American Mathematical Society at http://www.ams.org/bull/2007-44-04/S0273-0979-07-01168-8/S0273-0979-07-01168-8.pdf)

The Fields Medal is often called the Nobel Prize for mathematics, although it differs from the Nobel in several key ways. For one, the Fields is only awarded once every four years—although in recent years there's been a tendency to award it to four people each time. The second is that a recipient must be under the age of 40, and the selection committee hews to this: Andrew Wiles' proof of Fermat's last theorem was completed when he was slightly older than 40, and while he received a special medal and recognition, he did not receive the Fields Medal. Terence Tao is a 2006 Fields Medalist, and in this two-part article, he tackles an elusive question, "What is *good* mathematics?" His thoughts in the first part are quite interesting and accessible to all; in the second part, he illustrates some of his categories of "good math" via a case study of Szemerédi's theorem.

Patterns

Symmetry, by Herman Weyl.

Weyl was a mathematician who did a lot of work in physics, notably quantum mechanics. This classic book explores symmetry in nature and mathematics. Weyl once told Freeman Dyson, "My work always tried to unite the true with the beautiful, but when I had to choose one or the other, I usually chose the beautiful." I'm not sure they're words to live by, but I find them profound.

Number Theory

The Mathematics of Ciphers, by S. C. Coutinho.
Coutinho is a computer scientist in Brazil. The book consists of engaging expositions of primality, prime number testing, and the RSA cryptography scheme intended for a first-year class in computer science. The translated work is relatively easy to read and builds to some interesting ideas. Because it was slated for nonmathematicians, Coutinho's perspective is that of a keen-eyed outsider.

"It Is Easy to Determine Whether a Given Integer Is Prime," by Andrew Granville (*Bulletin of the American Mathematical Society* 42(2004): 3–38, available as a free .pdf download from the American Mathematical Society at http://www.ams.org/bull/2005-42-01/S0273-0979-04-01037-7/S0273-0979-04-01037-7.pdf)
This article summarizes and explains some of the huge breakthroughs that occurred in the search for "large" prime numbers after Agrawal, Kayal, and Saxena's paper "PRIMES is in P" appeared in 2004. By my highly subjective rating, although very much worth the effort, this is the hardest reading appearing on this list, and it probably requires the equivalent of an undergraduate degree in mathematics. Because this field is exploding, and because of the importance to e-commerce, I'd guess that all of these results have since been extended and refined, but still it's worth a look.

Real Analysis and Measure Theory

The Pea and the Sun: a Mathematical Paradox, by Leonard Wapner.
Wapner's book is pitched at the level of bright, mathematically inclined high school students who've (perhaps) heard of the Banach-Tarski paradox. This counterintuitive construction explains how to disassemble a small solid ball into a finite number of nonmeasurable sets, and then reassemble the pieces into a very large solid ball. Along the way, Wapner gets at some of the ideas of measure theory, and gives nice proofs that lead up to the main result. I liked this book a lot.

Measure and Category, by John C. Oxtoby.
This slender book is one of the publisher Springer-Verlag's infamous yellow "Graduate Texts in Mathematics." (Infamous among math

graduate students, at any rate.) Although it's probably necessary to have the equivalent of an undergraduate math education to profit from reading it, the writing is so light, clean, and lively, and the results are so enrapturing, I am pleased to recommend it.

Nonstandard Analysis

The Problems of Mathematics, by Ian Stewart.

Although Stewart's book encompasses many other nifty mathematical ideas, in particular it contains a chapter outlining the nuances and the history of some of the issues surrounding nonstandard analysis, including the subtle distinction between Leibniz's static infinitesimals and Newton's variable fluxions.

Non-standard Analysis, by Abraham Robinson.

Robinson's seminal work is for an enterprising individual with the equivalent of, say, a master's-level education in mathematics or logic.

Topology, Manifolds, and Cosmology

Beyond the Third Dimension: Geometry, Computer Graphics, and Higher Dimensions, by Thomas Banchoff.

Banchoff is a talented geometer and a masterful expositor. His book, from the Scientific American Library Series, is lavishly illustrated with beautiful images. The aim of the book is to encourage the visualization of higher dimensional spaces, and it can be fruitfully read by a bright high school student. It's a marvelous resource for anyone trying to think outside the (three-dimensional) box.

The Shape of Space, by Jeff Weeks.

Weeks has produced a luminous work, comparable to Oxtoby's *Measure and Category*, that takes advanced ideas and presents them so clearly and compellingly that it feels like everyone could and should read it. At any rate, if you were gripped by the Math Aftermath "Flat-Out Disoriented" and need more, Weeks is a good place to start.

A Homomorphism

Contemporary Abstract Algebra, by Joe Gallian.

Gallian's book is the friendliest introduction I've seen to algebraic groups and homomorphisms. Even so, it's aimed at sophomore- and junior-level math majors, and, as such, may require a large dollop of commitment.

Mathematical Foundations of Information Theory, by A. Ya. Khinchin.

Pretty technical, and redolent with the spare language of the professional mathematician, but I learned a lot about information theory from it when I was just beginning to be interested in mathematics. In particular, I found the discussions of entropy and information theory lucid and comprehensible.

An Introduction to Information Theory, by John R. Pierce.

I haven't read this, but I stumbled across it when I was looking up the exact title of Khinchin's book. It looks good and may be an easier read than Khinchin's book.

The Magical Maze: Seeing the World Through Mathematical Eyes, by Ian Stewart.

Another book by Stewart, a great explainer and popularizer of mathematics. This book, too, is filled with all sorts of good things; among them is a solid introduction to Turing machines.

Gödel, Escher, Bach: An Eternal Golden Braid, by Douglas Hofstadter.

What praise can I apply to this book that hasn't been already written? Winner of the Pulitzer Prize, it leads nonspecialists to the ideas of Gödel, reframes self-referentiality and paradox, and—I think—is the site of the first appearance of the evocative phrase "strange loop." It's marvelously witty, profoundly deep, and it heralded a new genre in letters.

Gödel's Theorem: An Incomplete Guide to Its Use and Abuse, by Torkel Franzén.

Franzén's book concisely achieves its goal of clearly demarcating the extent of applicability of Gödel's theorem. Along the way of accomplishing that, he shows its power and majesty in the fields of set theory and foundations, and brings into sharp focus many amusing nuances.

Gödel's Proof, by Ernst Nagel and James Newman.

Nagel and Newman take a dedicated reader step by step through a proof of Gödel's theorem. A classic.

Any Book by Raymond Smullyan, by Raymond Smullyan.

Actually, there is no such book by Smullyan (although I think he'd appreciate the self-referential title). He's an influential mathematical logician who, in addition to publishing serious works, has also written many highly readable books weaving together Gödel's theorem, truth, lies, formal systems, knights and knaves, islands, detectives, logic machines...

Critical Points

Contributions to the Founding of the Theory of Transfinite Numbers, by Georg Cantor.

Today's notation looks different, and the formulation of the ideas is simpler and cleaner, but it's wonderful to seize the opportunity to read, albeit in translation, the seminal work on infinity from the creator of set theory and the first imaginer of different sizes of infinity.

To Infinity and Beyond, by Eli Maor.

A user-friendly introduction to many forms of infinity, including Cantor's notions.

The Divine Proportion, by H. E. Huntley.

Huntley's work is an accessible reference which rewards a patient reader with a multiplicity of extrapolations from Fibonacci's sequence to the real world.

Openings

Mathematics and the Imagination, by Edward Kasner and James Newman.

A gift from Bioy Casares to Borges, this book is still surprisingly readable. Today, perhaps most notable for being the introduction of the word "googol" into the English language. It was coined by Kasner's nine-year-old nephew.

Fixed Points, by Yu. A. Shashkin.

Aimed at bright high-schoolers, this is the most elementary exposition of fixed points and Brouwer's theorem of which I am aware.

Borges y la matemática, by Guillermo Martinez.

Martinez is best known in America for his mystery *The Oxford Murders*, but he was first a mathematical logician. The essays are in Spanish, broadly cover Borges' writings, and begin with explanations of infinity that inform the story "The Aleph."

Bibliography

In sum, about all we can really say in paper's favor is that it's lighter than clay, less awkward than papyrus, and cheaper than parchment.
—Gregory Rawlins, *Moths to the Flame*

Abbott, Edwin A. *Flatland: A Romance of Many Dimensions*. 1884. Reprint, New York: Dover, 1992.

Adam, David. "Give six monkeys a computer, and what do you get? Certainly not the Bard." *The Guardian*, 9 May 2003. http://www.guardian.co.uk/arts/news/story/0,11711,952259,00.html (accessed 26 February 2008).

Alazraki, Jaime. *Borges and the Kaballah: and Other Essays on his Fiction and Poetry*. New York: Cambridge University Press, 1988.

Amaral, Pedro. "Borges, Babel y las matemáticas." *Revista Iberoamericana* 37(1971): 421–28.

Ammon, Ted. "A Note on a Note in 'The Library of Babel.'" *Romance Notes* 33(1993): 265–69.

Apostol, Tom. *Calculus, Volume II*. 2nd ed. New York: John Wiley & Sons, 1969.

Archimedes. *The Works of Archimedes*. Translated by T. L. Heath. New York: Dover Publications, 1950.

Babel, Isaac. *The Complete Works of Isaac Babel*. Edited by Nathalie Babel, translated by Peter Constantine. New York: Norton, 2002.

Banchoff, Thomas. *Beyond the Third Dimension: Geometry, Computer Graphics, and Higher Dimensions*. New York: Scientific Americian Library: Distributed by W. H. Freeman, 1990.

Barrenechea, Ana María. *Borges: the Labyrinth Maker*. Translated by Robert Lima. New York: New York University Press, 1965.

Barth, John. *Further Fridays*. Boston: Little, Brown and Co., 1995.

Bell-Villada, Gene. *Borges and His Fiction*. Rev. ed. Austin: University of Texas Press, 1999.

Benardete, José. *Infinity: An Essay in Metaphysics*. Oxford: Clarendon Press, 1964.

Bioy Casares, Adolfo, and Jorge Luis Borges. *Chronicles of Bustos Domecq*. Translated by Norman Thomas di Giovanni. New York: Dutton, 1976.

Bioy Casares, Adolfo, and Jorge Luis Borges. *Six Problems for Don Isidro Parodi*. Translated by Norman Thomas di Giovanni. New York: Dutton, 1981.

Borges, Jorge Luis, and Norman Thomas di Giovanni. *The Aleph and Other Stories*. New York: Dutton, 1978.

Borges, Jorge Luis. *Collected Fictions*. Translated by Andrew Hurley. New York: Viking, 1998.

―――. *Ficciones*. 1944. Rev. ed. 1956. Reprint: Buenos Aires: Emecé, 2001.

―――. *Ficciones*. Edited by Anthony Kerrigan, translated by Anthony Bonner. New York: Grove, 1999.

―――. *Labyrinths: Selected Stories and Other Writings*. Edited by Donald A. Yates and James E. Irby. London: The Folio Society, 2007.

―――. *Labyrinths: Selected Stories and Other Writings*. Edited by Donald A. Yates and James E. Irby. New York: New Directions, 1964.

―――. *The Library of Babel*. Translated by Andrew Hurley, illustrated by Erik Desmazieres. Boston: David R. Godine, 2000.

―――. *Selected Non-Fictions*. Edited by Eliot Weinberger. New York: Penguin Books, 1999.

Burger, Edward, and Michael Starbird. *The Heart of Mathematics*. Emeryville: Key College Publishing, 2005.

Burton, Robert. *The Anatomy of Melancholy*. Introduction by Holbrook Jackson. New York: New York Review Books Classics, 2001.

Campbell, Jeremy. *Grammatical Man: Information, Entropy, Language, and Life*. New York: Simon and Schuster, 1982.

Cantor, Georg. *Contributions to the Founding of the Theory of Transfinite Numbers*. New York: Dover Publications, 1952.

Capobianco, Michael. "Mathematics in the *Ficciones* of Jorge Luis Borges." *The International Fiction Review* 9(1982): 51–54.

Christensen, James J. *The Structure of an Atom*. London: Wiley, 1990.

Coutinho, S. C. *The Mathematics of Ciphers*. Natick: A. K. Peters, 1999.

Delsemme, Armand. *Our Cosmic Origins: From the Big Bang to the Emergence of Life and Intelligence*. Cambridge: Cambridge University Press, 1998.

Descartes, Rene. *Discourse on Method and the Meditations*. Translated by F. E. Sutcliffe. New York: Penguin Books, 1983.

Eco, Umberto. "Between La Mancha and Babel." *Variaciones Borges* 4(1997): 51–62.

Escher, M. C. *The Graphic Work*. Translated by John Brigham. Berlin: B. Taschen, 1990.

Ferrero, José María, and Alfredo Raúl Palacios. *Borges algunas veces matematiza*. Buenos Aires: Ediciones del 80, 1986.

Faucher, Kane X. "A Few Ruminations on Borges' Notions of Library and Metaphor." *Variaciones Borges* 12(2001): 126–37.

Foucault, Michel. *Discipline and Punish*. Translated by Alan Sheridan. New York: Pantheon, 1977.

Fraenkel, Abraham A. *Abstract Set Theory*. Amsterdam: North-Holland, 1953.

Franzén, Torkel. *Gödel's theorem: an incomplete guide to its uses and abuses*. Wellesley: A. K. Peters, 2005.

Gallian, Joseph. *Contemporary Abstract Algebra*. 5th ed. Boston: Houghton Mifflin, 2002.

Gamow, George. *One, Two, Three... Infinity*. 1947. Reprint, New York: Viking, 1954.

Grau, Christina. *Borges y la arquitectura*. Madrid: Cátedra, 1989.

Hausdorff, Felix. *Grundzüge der Mengenlehre*. 1914. Reprint, New York: Chelsea, 1949.

Hayles, N. Katherine. *The Cosmic Web: Scientific Field Models and Literary Strategies in the Twentieth Century*. Ithaca: Cornell University Press, 1984.

Hernández, Juan Antonio. *Biografía del infinito: la noción de transfinitud en George Cantor y su presencia en la prosa de Jorge Luis Borges*. Caracas: Comala.com, 2001.

Hofstadter, Douglas. *Gödel, Escher, Bach: an Eternal Golden Braid*. New York: Basic Books, 1979.

Huntley, H. E. *The Divine Proportion: A Study in Mathematical Beauty*. New York: Dover, 1970.

Kasner, Edward, and James Newman. *Mathematics and the Imagination*. New York: Simon and Schuster, 1940.

Keiser, Graciela. "Modernism/Postmodernism in 'The Library of Babel': Jorge Luis Borges's Fiction as Borderland." *Hispanófila* 115(1995): 39–48.

Keisler, H. Jerome. *Elementary Calculus: An Infinitesimal Approach*. Boston: Prindle, Weber & Schmidt, 1986.

Kesten, Hermann. *Copernicus and his World*. Translated by E. B. Ashton and Norbert Guterman. New York: Roy Publishers, 1945.

Khinchin, A. I. *Mathematical Foundations of Information Theory*. New York: Dover Publications, 1957.

Kirk, G. S., J. E. Raven, and M. Schofield. *The Presocratic Philosophers, A Critical History with a selection of texts*. 2nd ed. Cambridge: Cambridge University Press, 1995.

Körner, T. W. *The Pleasures of Counting*. Cambridge: Cambridge University Press, 2000.

Lasswitz, Kurd. "The Universal Library." Translated by Willy Ley. In *Fantasia Mathematica*, edited by Clifton Fadiman. New York: Copernicus, 1997.

Lofting, Hugh. *The Story of Doctor Dolittle.* 1920. Reprint, Philadelphia: Lippincott, 1948.

Luminet, Jean-Pierre, Glenn D. Starkman, and Jeffrey Weeks. "Is Space Finite?" *Scientific American,* April 1999: 90–97.

Maor, Eli. *To Infinity and Beyond.* Boston: Birkhäuser, 1987.

Martínez, Guillermo. *Borges y la matemática.* Buenos Aires: Editorial Seix Barral, 2006.

McKinzie, Mark, and Curtis D. Tuckey. "Higher Trigonometry, Hyperreal Numbers, and Euler's Analysis of Infinities." *Mathematics Magazine* 74(2001): 339–68.

Merrell, Floyd. *Unthinking Thinking: Borges, Mathematics and the New Physics.* West Lafayette: Purdue University Press, 1991.

Miller, Lynn F., and Sally S. Swenson, *Lives and Works: Talks with Women Artists.* Metuchen: Scarecrow Press, 1981.

Nagel, Ernst, and Newman, James. *Gödel's Proof.* New York: New York University Press, 2001.

Nicolas, Laurent. "Borges et l'infini." *Variaciones Borges* 7(1999): 88–146.

Østergaard, Svend. *The Mathematics of Meaning.* Aarhus: Aarhus University Press, 1997.

Oxford English Dictionary. 2nd ed. Oxford: Clarendon Press, 1989.

Oxtoby, John. *Measure and Category.* New York: Springer-Verlag, 1971.

Peters, John Durham. "Resemblance Made Absolutely Extact: Borges and Royce on Maps and Media." *Variaciones Borges* 25(2008): 1–24.

Pierce, John R. *An Introduction to Information Theory.* New York: Dover Publications, 1980.

Poincaré, Henri. *The Foundations of Science.* Translated by George Halstead. Lancaster: Science Press, 1946.

Rawlins, Gregory. *Moths to the Flame: The Seductions of Computer Technology.* Cambridge: MIT Press, 1996.

Robinson, Abraham. *Non-standard Analysis.* Rev. ed. Princeton: Princeton University Press, 1996.

Rodríguez Monegal, Emir. *Jorge Luis Borges: a Literary Biography.* Dutton: New York, 1978.

Rucker, Rudolph v. B. *Geometry, Relativity, and the Fourth Dimension.* New York: Dover, 1977.

———. *Infinity and the Mind.* Princeton: Princeton University Press, 1995.

Russell, Bertrand. *Human Knowledge: Its Scope and Limits.* London: Routledge, 1948.

———. *Principles of Mathematics.* Cambridge: Cambridge University Press, 1903.

Salpeter, Claudio. "La matemática biblioteca de Babel." Temakel. http://www.temakel.com/artborgesbabel.htm (accessed 26 February 2008).

Sarlo, Beatriz. *Jorge Luis Borges: A Writer on the Edge.* New York: Verso, 1993.

Shannon, Claude. "A mathematical theory of communication." *Bell System Technical Journal* 27(1948): 379–423, 623–56.

Shashkin, Yu. A. *Fixed Points*. Providence: American Mathematical Society, 1991.
Singh, Simon. *Fermat's Enigma: the Quest to Solve the World's Greatest Mathematical Problem*. New York: Walker, 1997.
Slusser, George. "Bookscapes: Science Fiction in the Library of Babel." In: *Mindscapes: The Geographies of Imagined Worlds*, edited by George Slusser and Eric Rabkin. Carbondale: Southern Illinois University Press, 1989.
Spencer-Brown, G. *Probability and Scientific Inference*. London: Longmans, Green and Co., 1957.
Stewart, Ian. *The Magical Maze: Seeing the World Through Mathematical Eyes*. New York: John Wiley & Sons, 1998.
———. *The Problems of Mathematics*. New York: Oxford University Press, 1987.
Sturrock, John. *Paper Tigers: The Ideal Fictions of Jorge Luis Borges*. Oxford: Clarendon Press, 1977.
Stoicheff, Peter. "Chaos of Metafiction." In *Chaos and Order: Complex Dynamics in Literature and Science*, edited by N. Katherine Hayles. Chicago: University of Chicago Press, 1991.
Tao, Terence. "What is good mathematics?" *Bulletin of the American Mathematical Society* 44(2007): 623–34.
Toca, Antonio F. "The Architecture of J. L. Borges." *Architecture and Urbanism* 123(1984): 104–07.
Wapner, Leonard. *The Pea and the Sun: A Mathematical Paradox*. Wellesley: A. K. Peters, 2005.
Weeks, Jeffrey. *The Shape of Space: How to Visualize Surfaces and Three-Dimensional Manifolds*. New York: M. Dekker, 1985.
Weyl, Herman. *Symmetry*. Princeton: Princeton University Press, 1952.
Wheelock, Carter. *The Mythmaker: A Study of Motif and Symbol in the Short Stories of Jorge Luis Borges*. Austin: University of Texas Press, 1969.
Williamson, Edwin. *Borges, A Life*. New York: Viking, 2004.
Woodall, James. *Borges: A Life*. New York: Basic Books, 1996.

Index

A set of ideas, a point of view, a frame of reference is in space only an intersection, the state of affairs at some given moment in the consciousness of one man or many men, but in time it has evolving form, virtually organic extension. In time ideas can be thought of as sprouting, growing, maturing, bringing forth seed and dying like plants.

—John Dos Passos, "The Use of the Past"

1-space, 166
1-sphere, 63, 166
2-Klein bottle, 87
2-manifold, 72, 76–78, 80
2-space, 60, 166
2-sphere, 62–64, 66, 68, 166
2-torus, 82
3-Klein bottle, 70, 86–87, 92, 166
3-space, 58–59, 166
3-sphere, 36, 63–64, 66–68, 84–85, 148–156, 166
3-torus, 70, 80, 82, 84–85, 92, 166
4-space, 166

Africa, 128
Air shafts, 99–100, 129
Analysis, xii
Annulus, 65, 166
Archimedean property, 132, 166

Archimedes, 11
Ars combinatoria, 26–29
Aurelius, Marcus, 107
Authorship, 32
Axioms
  definition of, xv, 167
  list of, 4–5

Babel, Isaac, xi
Base 10, 38
Base of an exponent, 167
Bilateral symmetry, 89
Blockhead, William Goldbloom, 127–128
Book-Man, 138
"Book of Sand," 45–46, 131
Books
  contents of, 34
  description of, 34–36

Books (*continued*)
distinct number of, 16–17, 21, 24, 107
first page of, 32–33
grain-of-sand, 19
"infinitely thin" pages, 46–54
lack of organization in distribution of, 36
number of, 12–13, 21
orderings of, 111–112, 114
shelving pattern of, 59
short description of, 34–36
spine of, 18
thickness of, 51–52
Brigg, Henry, 25
Brouwer's fixed point theorem, 163, 167
Burger, Edward B., 175–176
Burton, Robert, 141–142

Calculus, 53–54
Call numbers, 31
Cantor, Georg, 145–146
Catalogue
description of, 30–31
false, 32
form of, 31
Library as, 35, 39
Catalogue card, 30–31
Cavalieri's principle, 48, 167
Center of a sphere, 61, 69–70
Central points, 59
Chartes Cathedral, xvii–xviii
Chiliagon
definition of, 167
Descartes' thousand-sided, xx
Church–Turing thesis, 124
Cipher, 139
Circle
codimension of, 77, 167
description of, 77
Pythagorean theorem used to derive equations for, 150–155
Circumference, 68, 85, 167

Closed interval, 48–49, 167
Codimension, 77, 167
Combinatorial analysis, 107
Combinatorics, 13–14, 107–119, 168
Computers, 121–122
Constructivists, 119
Cosmology, 57, 178
Countable, 162, 168
Coutinho, S.C., 177
Critical points, 126–140, 179–180

DeGroot, Morris, 72
Denominator, 27, 168
Descartes, Rene
exponential notation, 11
origin, 59
thousand-sided chiliagon, xx
Description of books, 34–36
Distinct books, 16–17, 21, 24, 107

Eco, Umberto, xi, 16
Elizabeth I, 159
Empty set, 168
Endpoints, 83
Escher, M. C., xvii
Euclid, 40–42
Euclidean plane, 62, 64, 76, 151
Euclidean space, 58–59, 62, 168
Euler, Leonhard, 43, 110
Exponent
negative, 13
positive integer, 12
Exponential notation, 11, 168

Factorial
definition of, 109–110, 168
description of, 109–110
Stirling's approximation to the, 111, 173
Factorizations, 40–41
Factor (noun), 168
Factor (verb), 168
False catalogues, 32

Fibonacci sequence, 137–138, 168–169
First pages of books, 32–33
Fixed point, 169
Fixed point theorem, 163
Flat, 169
Floor plans, 97–98, 101–103
Fluxions, 53
Four-dimensional Euclidean space, 63, 72
Fractals, 144, 169
Franzén, Torkel, 179
Function, 23, 169
Funes-like, 116, 169

Gallian, Joe, 178
Gamma function, 110, 169
"Garden of the Forking Paths, The," 68
Gödel, Kurt, 121
Goethe, Johann Wolfgang von, 107
Googol, 43
Grain-of-sand books, 19
Grand Pattern
    description of, 107–108, 113–115
    Other Entity's, 115–116, 118
Granville, Andrew, 177
Gravitational field, 161
Great circles, 68–69, 169

Hadamard, 43
Hayles, N. Katherine, 132–136
Hexagon(s)
    adjacent, 102–103
    air shaft through, 99–100, 129
    capacity of, 42
    as clones, 99
    description of, 22, 36–37, 41
    doorways to, 103
    sides of, 94
    spiral staircase with, 96, 99, 102
    square vs., 92
Hexagonal prism, 91, 169
Hofstadter, Douglas, 179

Homomorphism, 120–125, 169, 178–179
Hopf fibration, 161
Huntley, H.E., 180
Hyperreal number line, 54, 170
Hyperreal numbers, 54, 169

Imbert, Stefano, 93
Incompleteness theorems, 121
"Infinitely thin," 46–54
Infinite set, 162
Infinitesimals, 53–54, 170
Infinite sums, 46
Infinite union of intervals, 52
Information theory, 30–44
Initial position, 123, 170
Integers
    condensed form of, 37
    definition of, 170
    positive. *See* Positive integers
Internal state, 123, 170
Irrational numbers, 162, 170

Kasner, Edward, 180
Khinchin, A. Ya, 178
k integers, 28
Klein bottle
    2-, 87
    3-, 70, 86–87, 92, 166
    black flag in, 78–79
    codimension, 77
    definition of, 73, 170
    disorientation involving, 78–79
    four-dimensional representation of, 74
    Library modeled on, 73–76
    three-dimensional representation of, 74
    universe as, 78
Koch snowflake curve, 144, 170

Labyrinth, xvii–xviii
Lasswitz, Kurd, 127
Lebesgue, Henri, 48

Leibniz, 53
Lemma, 102, 170
Libits
    abutting, 115
    definition of, 111–112, 170
    shape of, 117–118
"Library of Babel"
    artistic renditions of, 93
    books in. *See* Books
    design of, 93–94
    global order of, 136
    hexagons in. *See* Hexagons
    origin of, 130
    pages in, 46
    path through, 124
    shape of, 108
    size of, 22
    structure of, 101
    testing of, 85
Library unit, 170
Line
    endpoints on, 83
    hyperreal number, 54, 169
    real number, 48–49, 54, 172
Locally Euclidean, 60, 170
Log(10), 25
Logarithm
    definition of, 171
    notation for, 23
    solving equations using, 24–25, 55–56
Logarithmic function, 19
Lower bound, 13, 171
Luxemburg, Wilhelmus, 53

Manifold
    2-, 72, 76–78, 80
    definition of, 171
    description of, 60–61, 71
    readings about, 178
Maor, Eli, 180
Map, 120, 171
Martínez, Guillermo, 180
Measure 0, 49, 51, 173
Measure theory, 177

Median, 37, 43, 171
Menard, Pierre, 32
Merrell, Floyd, 132, 136–137
Miguel Cané Municipal Library, 30, 94–95
Mirror-reflection, 79–80
Mirror-reversals, 89–90, 136
Mirrors, 85–86, 88
Möbius band, 75, 136, 171

Nagel, Ernst, 179
Natural log function, 43–44
Negative exponent, 13
Newman, James, 179, 180
Newton, Isaac, 53
Non-Euclidean, 60, 171
Nonorientable space, 90, 171
Nonstandard analysis, 53, 171, 178
Non-standard Analysis, 53, 178
Notations, 157
Numbers
    call, 31
    hyperreal, 54, 169
    irrational, 163, 170
    prime, 40, 172
    real, 172
Number theory, 177
Numerator, 172

One-sided 2-manifold, 78
One-to-one correspondence, 172
Orderings of books, 111–112, 114
Origin, 59, 172
Orthographic symbols, 14, 16, 28, 38, 88, 98
Østergaard, Svend, 131
Other Entity, 115–116, 118
Output number, 23
Oxtoby, John C., 177–178

Pages
    description of, 46
    "infinitely thin," 46–54
    Parmenides's paradox, 46

"Pascal's Sphere," 57–58
Peirce, Charles Sanders, 165
Perception of space, 4
Periodic, 68, 172
Pierce, John R., 179
Poincaré, Henri, 144–145
Point, 60–61
Positive integers
    base 10 used to represent, 38
    exponent, 12
    unique factorization theorem, 40
Power of 10
    definition of, 172
    description of, 18
Prime number, 40, 172
Prime number theorem, 43
Product, 40, 172
Punctuation, 160
Pythagorean theorem, 150

Raised to a power, 172
Rational numbers, 51, 172
Readings, 175–180
Real analysis, 45–56, 177–178
Real number line, 48–49, 54, 172
Real numbers, 172
Repetitions, 28
Roberts, Joe, 27
Rotation, 79–80
Royce, Josiah, 163
Russell, Bertrand, 143, 145–146

Sarlo, Beatriz, 95–96, 128–129
Scientific notation, 11
Self-referentiality, 31
Set
    definition of, 28, 172
    diameter of, 173
    infinite, 162
Set of measure 0, 49, 173
Set theory, 173
Shakespeare, William, 159
Shannon, Claude, 30

Shashkin, Yu. A., 180
Sight lines, 129
Smullyan, Raymond, 179
South Africa, 128
Space
    1-, 166
    2-, 60, 166
    3-, 58–59, 166
    definition of, 173
    perception of, 4
Sphere
    1-, 63, 166
    2-, 62–64, 66, 68, 166
    3-, 36, 63–64, 66–68, 84–85, 148–156, 166
    center of, 61, 69–70
    Pythagorean theorem used to derive equations for, 150–155
Spinal letters, 18
Spiral staircases, 88, 93–94, 96, 98–99, 102
Square, 72, 92
Starbird, Michael, 175–176
"Stark and depressing conclusion," 102–105
Stewart, Ian, 178, 179
Stirling's approximation to the factorial, 111, 173
"Strange Loop," 133
Subsets, 28, 173

Tao, Terence, 176
Theory of infinite sums, 46
Tiling of space, 111, 173
"Tlön, Uqbar, Orbis Tertius," 159
"Top-front" labeling, 89
Topology, 57, 173, 178
Torus
    2-, 82
    3-, 70, 80, 82, 84–85, 92, 166
    definition of, 71, 83, 92, 173
    flat property of, 72
    Library modeled on, 71–73
    square and, 72

Transfinite numbers, 145, 173
Turing, Alan, 121–122
Turing machine, 122–124
Twin primes, 122

Uncountable, 162, 173
Union of intervals, 49, 52
Unique factorization theorem, 40–42, 173
Universe
   3-Klein bottle as model for, 86–87
   3-sphere as model for, 84
   3-torus as model for, 84
   description of, 3–4
   grains of sand, 20
   as Klein bottle, 78
   outside of, 61
   size estimations, 19

Upper bound
   creation of, 11
   definition of, 174

Venn diagram
   definition of, 174
   illustration of, xii
Vindications, 32, 106
Volumes
   distinct number of, 16–17, 21, 24
   scattering of, 31

Wapner, Leonard, 177
Weeks, Jeff, 91, 178
Well-ordering principle, 118–119, 174
Weyl, Herman, 176–177
Wiles, Andrew, xiii
Wittgenstein, Ludwig, 35–36

Zeno's Paradox, 46–47, 51, 146

Yo conozco distritos en que los jóvenes se prosternan ante los libros y besan con barbarie las páginas, pero no saben descifrar una sola letra. Las epidemias, las discordias heréticas, las peregrinaciones que inevitablemente degeneran en bandolerismo, han diezmado la población. Creo haber mencionado los suicidios, cada año más frecuentes. Quizá me engañan la vejez y el temor, pero sospecho que la especie humana — la única — está por extinguirse y que la Biblioteca perdurará: iluminada, solitaria, infinita, perfectamente inmóvil, armada de volúmenes preciosos, inútil, incorruptible, secreta.

Acabo de escribir infinita. No he interpolado ese adjetivo por una costumbre retórica, digo que no es ilógico pensar que el mundo es infinito. Quienes lo juzgan limitado, postulan que en lugares remotos los corredores y escaleras y hexágonos pueden, inconcebiblemente cesarlo cual es absurdo. Quienes lo imaginan sin límites, olvidan que los tiene el número posible de libros. Yo me atrevo a insinuar esta solución del antiguo problema: <u>La Biblioteca es ilimitada y periódica</u>. Si un eterno viajero la atravesara en cualquier dirección, comprobaría al cabo de los siglos que los mismos volúmenes se repiten en el mismo desorden (que, repetido, sería un orden: el Orden). Mi soledad se alegra con esa elegante esperanza.

Jorge Luis Borges

Printed in the USA/Agawam, MA
March 27, 2023

807564.058